Nomos Universitätsschriften

Betriebswirtschaftslehre

Volume 5

Nils Niehues

An Agency Perspective on Voluntary CO2 Disclosure

A Mixed-Method Study

Nomos

The Deutsche Nationalbibliothek lists this publication in the
Deutsche Nationalbibliografie; detailed bibliographic data
are available on the Internet at http://dnb.d-nb.de

a.t..: Siegen, Univ., Diss., 2018

ISBN   978-3-8487-5049-8 (Print)
        978-3-8452-9221-2 (ePDF)

British Library Cataloguing-in-Publication Data
A catalogue record for this book is available from the British Library.

ISBN   978-3-8487-5049-8 (Print)
        978-3-8452-9221-2 (ePDF)

Library of Congress Cataloging-in-Publication Data
Niehues, Nils
An Agency Perspective on Voluntary CO2 Disclosure
A Mixed-Method Study
Nils Niehues
210 p.
Includes bibliographic references.

ISBN   978-3-8487-5049-8 (Print)
        978-3-8452-9221-2 (ePDF)

1st Edition 2018
© Nomos Verlagsgesellschaft, Baden-Baden, Germany 2018. Printed and bound in
Germany.

Preface

In recent time, more and more companies expand their annual disclosures by adding carbon emission numbers or even notes about management approaches to lower the carbon emission impact on the environment. Although such a voluntary disclosure is very welcome, it seems rather unlikely that all companies report in the same style making comparisons possible. More generally, there are still a lot of unanswered questions about calculating complete carbon footprints and developing useful key performance indicators. Hence, identifying carbon efficient companies is a complex task for anyone, in particular for stakeholders and investors.

This study aims to develop a better understanding of the challenges and prerequisites that should be taken into account for analysing carbon emission disclosures. Starting with an overview about the regulatory frameworks for carbon emissions, different reporting schemes and assurance standards, Nils Niehues derives some interesting insights about the shortcomings of prior studies in the field (e.g. studies about the business case for CSR). The following empirical analyses shed some further light on the limited quality of recent carbon reports. The study evaluates the CDP data set over the reporting periods 2010 until 2015 and combines the results with indications from several expert interviews. Therefore, the results of the study allow different implications for regulators, standard setters, analysts and companies to improve the reporting quality in general and to enhance the ability of stakeholders identifying real carbon efficient companies.

Finally, I do hope that readers of this study will awaken their interest in the field of CSR reporting pitfalls and ways to identify real carbon efficient companies.

Andreas Dutzi
Siegen, March 2018

Acknowledgement

I would like to express my sincere gratitude to my advisor Prof. Andreas Dutzi for the continuous support of my Ph.D. study and related research, for his patience, motivation, and immense knowledge. His guidance helped me throughout the research and writing of this thesis. I would like to thank Prof. Volker Stein for serving as the second reader.

My sincere thanks also go to my DP DHL colleagues Patric Pütz, Dr. Klaus Hufschlag, and Andreas Schütt for all their encouragement during the past years. Without their precious support, it would not have been possible to conduct this research. I thank my fellow Ph.D. students for the stimulating discussions and for all the fun we have had in the last years. I would like to highlight here Dr. Bastian Rausch, Guido Kaufmann, and Daniel Ludwig. Representatively for all the feedback I received, I would like to thank my lector Steve Britt.

Last but not least, I would like to thank my parents Dorothea and Bernhard, my sister Sarah and my grandmother Helene for supporting me spiritually throughout writing this thesis and my life in general.

Nils Niehues
Bonn, March 2018

For Delia

Contents

List of Figures

List of Tables

List of Abbreviations

AA1000AS	AccountAbility 1000 Assurance Standard
BICS	Bloomberg Industry Classification Systems
BS	Balance Sheet
CC	CDP Questions on Climate Change
CDP	CDP, formerly Carbon Disclosure Project
CDSB	Carbon Disclosure Standard Board
CEO	Chief executive officer
CEP	Corporate Environmental Performance
CFP	Corporate Financial Performance
CO_2	Carbon Dioxide
CO_2e	Carbon Dioxide equivalent
CSI	Carbon stabilization intensity
CSO	Center for Sustainable Organizations
CSR	Corporate Social Responsibility
DAX	Deutsche Aktienindex
DIN	Deutsche Industrie Norm
DJSI	Dow Jones Sustainability Index
DMA	Disclosure Management Assessment
DRS	Deutsche Rechnungslegungsstandards
EBIT	Earnings before interest and tax
ESG	Environmental Social and Governance
ETS	Emission Trading System
EU	European Union
FTE	Full Time Employee
FTSE	Financial Times Stock Exchange
GEVA	Greenhouse gas emissions per unit of value added
GHG	Greenhouse Gas
GHGP	Greenhouse Gas Protocol
GICS	Global Industry Classification Standard
GLEC	Global Logistic Council
GRI	Global Reporting Initiative
GWP	Global Warming Potential
IAASB	International Auditing and Assurance Standards Board

ICLEI	International Council for Local Environmental Initiatives
IDW	Institut der Wirtschaftspruefer
IETA	International Emissions Trading Association
IFRS	International Financial Reporting Standards
IIRC	International Integrated Reporting Council
IPCC	Intergovernmental Panel on Climate Change
IR	Investor relations
ISAE	International Standard on Assurance Engagement
ISO	International Organization for Standardization
IT	Information technology
KLD	KLD Research & Analytics, Inc.
KPI	Key performance indicator
LCA	Life cycle assessment
MSCI	formerly Morgan Stanley Capital International
NGO	Non-governmental organization
OECD	Organisation for Economic Co-operation and Development
OLS	Ordinary Least Squares
P & L	Profit and Loss
PwC	PricewaterhouseCoopers
QUAL	Qualitative
QUAN	Quantitative
TCR	Transient climate response
TNO	Nederlandse Organisatie voor Toegepast
TRI	Toxic release inventory
UK	United Kingdom
UN	United Nations
USA	United States of America
WBCSD	World Business Council for Sustainable Development
WEF	World Economic Forum
WRI	World Resources Institute
WWF	World Wide Fund for Nature

Chapter 1: Introduction

A. *Relevance of climate change*

There is a huge research stream regarding climate change and its negative implications for human life on earth (IPCC, 2014; Stern, 2006; Pope Franscesco, 2015). The scientific community considers it self-evident that climate change is caused by human activity (IPCC, 2014). The impact of global warming extends beyond the destruction of nature (Marshall, 2009); it is also expected to have a significant negative financial impact on all economies worldwide (OECD, 2015). Additionally, and in part as a result, climate change has the potential to destabilize our society (Pope Franscesco, 2015, pp. 53 ff.).
Minimizing or at least mitigating the effects of global warming is thus the most important task for the entire population of the earth. All research disciplines are called upon to develop solutions to mitigate climate change.

Even though governments still subsidize carbon-intense energy exploration (Robertson and Adani, 2016; Porter, 2016), there are also significant approaches by regulators, communities and companies to develop actions to reduce the impact of climate change, either by reducing the root causes of climate change or taking mitigation action on the expected changes caused by climate change (Andrew and Cortese, 2013). This happens often on a global scale with the understanding of breaking down climate targets to lower, more applicable levels (CDP, 2015d; Kyo, 2014; Kyoto-Protokoll, 1997).

The current understanding is that climate change is caused by the combustion of fossil fuels, producing greenhouse gases (GHG). For purposes of simplification, communication about GHG often groups the seven gases together and describes them as carbon dioxide equivalent (CO_2e) or carbon dioxide (CO_2). This can be done by calculating GHG emissions to CO_2e based on their global warming potential (GWP). International regulators use these CO_2e figures to agree upon climate targets to limit the output of GHG/CO_2e emissions over a certain timeframe (Kyoto-Protokoll, 1997; Kyo, 2014; Uni, 2015). This regulatory force is one of several triggers that prompt companies to develop climate-change strategies and be transparent about the company's impact on climate change. Companies' climate-change activities are often embedded in the broader concept of Corporate Social Responsibility (CSR).

B. The business case for Corporate Social Responsibility

Reasons for implementing CSR measures are often discussed under the BUSINESS CASE FOR CSR (Carroll, 2010, pp. 85-105, Carroll, 2011, pp. 1-4). The arguments for this business case can be classified several ways. Two common ways are to structure the motivations into competitiveness, legitimation and ecological responsibility (Bansal and Roth, 2000, pp. 717-736). The arguments can also be structured into pro-active factors (motivations) and reactions to external pressure (drivers), as Okereke does for the CSR sub-category of climate actions (Okereke, 2007, pp. 479-482). The Table 1.1 shows that a company can choose climate actions based on own MOTIVATION or can refuse to act until there is a critical amount of pressure forcing the company to react.

One argument for managing carbon emissions is the increase in profit (Okereke, 2007, pp. 479-482, Niehues, 2014, pp. 265-273). This increase in profits results either through reduced costs (e.g. for energy) or increased revenue / margin based on respective products. Additionally, the company's credibility relative to climate change improves, and future external pressures are reduced. Another argument is described as a fiduciary obligation for the company's future: ensuring that the company adapts early to the changing environment to avoid subsequent losses due to market shifts and potential future regulations. Added to this are the physical risks resulting from climate change. E.g. insurance companies see their business models affected through climate change, or companies located near coasts want to contribute their share towards reducing the risk of a flood. The last motivation argument revolves around ethical considerations, especially from the upper echelon within the company (Okereke, 2007, pp. 479-482). On the other hand, even if companies do not have strong enough arguments to perform carbon management, there are certain external pressures that drive companies to manage their emissions.

High volatility of energy prices and the respective volatility impact on a company's profit and loss is one DRIVER for managing energy consumption and thus carbon emissions.

Changes in the market, e.g. due to customer demand for sustainable goods, can also persuade companies to change their supply chain and enforce carbon

Motivations	Profit
	Credibility and leverage in climate policy development
	Fiduciary obligation
	Guiding risk
	Ethical considerations
Drivers	Energy prices
	Market shifts
	Regulation and governments directives
	Investor pressure
	Technological Change
Barriers	Lack of strong policy framework
	Uncertainty about government's action
	Uncertainty about the marketplace

Table 1.1: Summary of Motivations, Drivers and Barriers Related to Corporate Climate Actions (Okereke, 2007, p. 481).

efficient actions. Other customer-driven forces are that the products of companies with responsible corporate behaviour might be perceived as products with higher value and able to be positioned with higher margins (Balderjahn, 2013).

The final argument for managing carbon emissions actively is technological change. With a changing cost and production environment, technological change can pressure companies into investing in energy-efficient technology, processes or business models. This general shift in strategy can support the disclosure of the respective (reduced) emissions, to prove to stakeholders, and to investors in particular, that environmental change is being taken into consideration (Okereke, 2007, pp. 482-483).

BARRIERS to implementation of carbon management relate to uncertainty. These uncertainties are the result of stakeholder requirements, uncertainty or limited expectations of financial benefit, or uncertainty about future regulation/marketplace conditions (Okereke, 2007, p. 481). To streamline the requirements due to future regulation and stakeholder expectations, to give guidance to companies as to indicate the areas in which companies need to take action to reduce their climate-risk exposure, several voluntary initiatives were founded. As motivations, drivers, and barriers evolve from interaction with stakeholders, disclosure of green actions is one way to interact with the stakeholders of the reporting company.

C.   *CO₂ emission data as proxy for environmental performance*

In general, the ability to gain extra rents based on environment-friendly behaviour is a function of market efficiency. Relying on agency theory, managers should send a signal to shareholders to reduce information asymmetry. A valid signal could be a carbon report to prove that the company is carbon-efficient (Niehues and Dutzi, 2018).

Therefore, several research papers use CO_2 efficiency or other CO_2 disclosure elements to operationalize green performance. This research stream is often considered as *pays-to-be-green literature* (Hahn et al., 2015). Additionally, companies face an increasing disclosure demand from investment communities to inform their investors about their environmental performance, especially their CO_2 emission inventory (UNG, 2015). After all, investors have a huge influence on what information companies disclose voluntarily (Armstrong, 2011, pp. 29-32, Wegener, 2013, p. 53). This trend accompanies the increasing importance of non-financial KPIs (Bundesministerium der Justiz, 2012; Hufschlag and Pütz, 2013). Publication of relevant non-financial KPIs can reduce the risk of insider trading and thus reduce agency conflicts (Luo, 2014, pp. 10, 28-30, Skelton, 2013, p. 420 ff.).

Additionally, increased climate regulation not only forces companies to manage their emissions but also changes carbon disclosure into a compliance topic for companies (Eleftheriadis, 2014, p. 10, Pattberg, 2012, pp. 615-618).

The purpose of this thesis in the field of accounting research is to evaluate the current environmental accounting practices in the area of carbon disclosure. This thesis evaluates the extent to which stakeholders can use carbon disclosure to identify carbon-efficient companies.

D.   *Research design - Exploratory sequential design*

To identify the possibilities with regard to the extent to which a voluntarily reported carbon number can be interpreted as a carbon efficient market signal, it is important to understand how this number is used in reality. This requires an exploratory research design with a constructivist worldview. Constructive worldviews aim to understand the *world out there* by including multiple views and generate/develop theory based on these views. As the aim is to understand if agency theory can explain high-quality voluntary disclosure, exploratory research designs in qualitative settings can be used to test the theoretical construct (Creswell, 2014, pp. 6-9). As carbon numbers are usually collected in big databases (e.g. CDP, Bloomberg, etc.) and analyzed in a quantitative setting, it is likely that the intended audience for this study will have a quantitative bias. It is thus important to support the outcome with quantitative results that underpin the findings obtained. This increases the acceptance of the qualitative results and analyzes if the findings can be generalized (Watkins and Gioia, 2015, p. 34). So a research design must be set up that includes qualitative and quantitative aspects to combine the strengths of both areas and minimize their respective limitations. Research designs that combine quantitative and qualitative views are often summarized under *mixed-methods approaches* (Creswell, 2014; Watkins and Gioia, 2015). Alternative terms for this include *multiple ways of hearing* (Greene, 2007, p. 20), *third research paradigma* (Johnson and Onwuegbuzie, 2004, p. 15), or *third methodological movement* (Teddlie and Tashakkori, 2003, p. 5). Mixed methods offer different options to mix the respective quantitative and qualitative research streams (Creswell and Clark, 2011; Creswell, 2014; Kuckartz, 2014; Watkins and Gioia, 2015). For a research problem in which the measurement model and the theoretical design have to be developed, first, it is important that the design is emergent, as expert information should be directly used to adjust the further steps. The recommended setting is an exploratory sequential design. As Figure 1.1 shows, research starts with qualitative data collection and data analysis. The results of this study define what data need to be collected in the quantitative study, which is then analysed

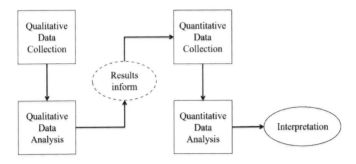

Figure 1.1: Exploratory sequential design (Watkins and Gioia, 2015, p. 33).

in a quantitative setting. The results from both studies can be used for the mixed-methods interpretation (Watkins and Gioia, 2015, pp. 26-34). The worldviews associated with this mixed-method approach change over time. First of all, it starts with the pragmatic worldview looking at the real-life problem of heterogeneous results in the literature review. The pragmatic worldview focusses on the problem and decides on methods afterwards, which is typical for the mixed-method researcher (Kuckartz, 2014). Based on this pragmatic worldview, the research setting of exploratory sequential design is chosen.

The qualitative setting follows a clear constructivist-worldview approach. Constructivism is usually connected to qualitative settings and aims to understand the complexity of the problem from different angles. Accordingly, the qualitative study samples carbon experts from different areas. The quantitative analysis follows a postpositivsm worldview. Under this worldview, the world is ruled by rules and cause effects. These are modeled in models that cover a reduced reality, and hypotheses are tested [Creswell, 2014, pp. 6-9, Watkins, 2015, pp. 32-34].

The translation of this research design into a research strategy can be summarized in Figure 1.2. This figure shows the preliminary work for this study, with its pragmatic worldviews; the qualitative part, in which data are collected through expert interviews; and the quantitative part, in which the data are gleaned from databases identical to information also available to investors and other stakeholders. Several authors also highlight the importance of proper notation here *qual-QUAN*. The *qual* stands for the qualitative part. The lower-case letters indicate that the qualitative study is not the focus. The arrow stands for a sequential design, while the dominating quantitative study is indicated by the *QUAN* (Watkins and Gioia, 2015, p. 33). The target audience of standard-setters and investors with their empirical focus explains the higher weighting of the quantitative part. This approach is consistent with some opinions of empirically focussed researchers that qualitative studies are *only useful to support empirical evidence* (Barton and Lazarsfeld (1979) quoted by Lamnek (2010)). At the same time, it is intended that some of the empirical tests will be used by the CDP, which forms the basis for several quantitative research efforts and thus has a direct impact on future CO_2 accounting research (Niehues, 2016, p. 12-13).

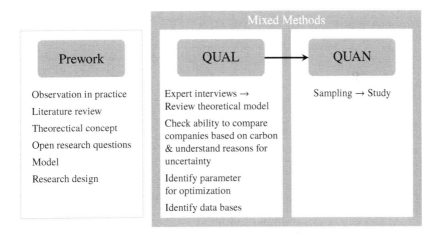

Figure 1.2: Visual model of the research strategy.

Chapter 2: Carbon Disclosure Framework

A. *Corporate Social Responsibility disclosure*

The area of CSR has evolved based on the theoretical foundation that companies need to satisfy stakeholder expectations (Parmar et al., 2010). This theoretical approach known as STAKEHOLDER THEORY adapts on the assumption that a corporation has to stay in dialogue with various stakeholders and incorporate their needs in the company's value creation process. This is required to achieve financial profit. Working in a network of stakeholders from which the company gains benefit/on which the company is dependent, the company has to work within their interests (Freeman and Liedtka, 1991, p. 96). The incorporation of stakeholder interests in strategic management was most prominently named by Freeman (1984). Figure 2.1 shows that the value creation model of a simplified input-output model (left) is extended by dialogues with all relevant stakeholders (right). This implies two different key changes to the input-output model. First, the landscape of a company's relevant stakeholders is extended to various stakeholders. Second, the interaction between stakeholders changes from one-way to dialogues. Managers and researchers should thus take this more complicated environmental setting of the company into account if decisions or analyses are undertaken. Interactions between stakeholders are ignored in this model to reduce complexity, even though these interactions exists (Mitchell et al., 1997, pp. 879-882).

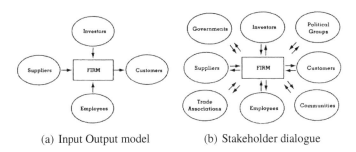

(a) Input Output model   (b) Stakeholder dialogue

Figure 2.1: Inclusion of the stakeholder

Based on this broader view of the environment in which companies act, theories arise based on their stakeholder needs. These theories come from the stakeholder setting analyzed. Looking at the different stakeholder settings, it is important to understand which stakeholders should be included. Here, Freeman (Freeman, 1984, p. 46) uses a rather broad definition of *any group or individual who can affect or is affected by the achievement of the organization's objectives.* Other researchers connect this question with the requirement of adopting stakeholder interests in the decision-making process and rank/cluster stakeholders based on stakeholder power, legitimacy and urgency. The analyses of the STAKEHOLDER SALIENCE [Guenther, 2015, p. 4, Mitchell, 1997, p. 856] help managers understand interests and effects and make these visible to managers in a structured way (Mitchell et al., 1997, pp. 855-868).

Alternative studies cluster stakeholders, based on their importance to the company, into primary and secondary stakeholders. Primary stakeholders are all groups relevant for the existence of the company, as these stakeholders control critical resources (Frooman, 1999, pp. 191-205, Guenther, 2015, p. 4, Pfeffer, 1978, pp. 521-532), and secondary stakeholders refer to the broad stakeholder definition of Freeman minus the primary ones (Buchholtz, 2012, p. 60, Gibson, 2000, p. 245, Parmar, 2010, p. 6). No matter how stakeholders are clustered, in all cases managers face expectations of their stakeholders which the company is dependent on (Parmar, 2010, p. 6, Rodgers, 2004, pp. 349-363, Scholes, 1998, pp. 227-238, Sharma, 2005, pp. 159-180, Wright, 1997, pp. 77-90).

Carroll situated this responsiveness to stakeholder requirements in a triple-bottom-line approach consisting of legal, ethical and financial targets (Carroll, 1979, p. 503). This approach was further developed by Elkington into a so-called 'triple bottom line' consisting of social, environmental and financial performance which are considered as the areas a company has to perform on to satisfy their stakeholder needs (Elkington, 1997, 2001).

The concept of CSR evolved from the need to satisfy stakeholder needs to actively include those targets into the core business strategy. The underlying assumption here is, that companies which actively manage those stakeholder expectations are able to also benefit stronger from these stakeholders. The inclusion of CSR targets into the business strategy was most prominently supported by Porter (Porter, 1987; Porter and Kramer, 2006; Porter et al., 1995). To gain a common understanding of the definition of CSR, Dahlsrud undertook research comparing 37 definitions of CSR. As most definitions for CSR have similar contents (Carroll, 1999, pp. 268-295, Dahlsrud, 2008,

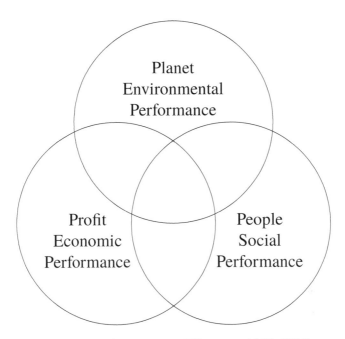

Figure 2.2: Triple bottom line concept (Elkington, 1997, 2001).

pp. 1-13), this document uses the EU definition, which in turn is built upon several definitions. These contain the United Nations Global Compact, the United Nations Guiding Principles on Business and Human Rights, the ISO 26000 Guidance Standard on Social Responsibility, the International Labor Organization Tripartite Declaration of Principles concerning Multinational Enterprises on Social Policy, and the OECD Guidelines for Multinational Enterprises (European Commission, 2007). The EU definition is as follows:

> Companies can become socially responsible by following the law (and) integrating social, environmental, ethical, consumer, and human rights concerns into their business strategy and operations (European Commission, 2007).

As this document focusses on the disclosure on GHG emissions, the EU strategy on CSR implies improvement of company disclosure of social and environmental information (European Commission, 2007). CSR as Corporate Social Responsibility is often operationalized into measurable KPIs. These KPIs are usually grouped into environmental, social and governance (ESG) indicators. Data-focussed analysis thus often uses ESG indicators similar to CSR.

As the causes and impacts of climate change are happening on a global scale, the solutions developed to tackle this issue must be global as well. This is why coalitions for climate change are often driven by powerful initiatives from globally active investors, multinational companies and political interest groups such as the World Economic Forum (WEF). With their power, voluntary initiatives strive to impact national regulations and the way companies, investors and other stakeholders share and perceive information about environmental impacts.

As there are endless initiatives that help companies improve their environmental footprint, the aim of this document is not to give full transparency to stakeholder initiatives, but rather explain some of the most relevant global initiatives which are often the basis for national specification. These are the Global Reporting Initiative (GRI), the World Resource Institute (WRI) and the World Business Council for Sustainable Development (WBCSD) which together publish the Greenhouse Gas Protocol (GHGP), the CDP (formerly Carbon Disclosure Project) and the International Integrated Reporting Council (IIRC).

As overarching framework for non financial disclosure the GRI and the IIRC are described. The GRI is an independent global non-profit non governmental organisation (NGO) located in Amsterdam with the mission of enabling companies to report on their economic, environmental, social and governance performance. The core outcomes of the GRI are the GRI standards. These standards are freely available. The standards are written in a multi-stakeholder process involving thousand of professionals from different sectors, constitutions and regions (GRI, 2013, p. 2).

As shown in Figure 2.3, the GRI standard is the overall standard for companies that seek to adopt CSR reporting. Different contents of the GRI are further specified, as the GHGP for example explains how carbon reporting is done. At the next level, there are several data collection databases (e.g. CDP) to enable stakeholders to get comparative data for CSR reporting across companies and time. These databases often link to other databases, such as Bloomberg terminal, to reach a wider community or facilitate access for certain stakeholder groups. As the GRI dominates the CSR reporting, the GRI standard is described in more detail.

The latest reporting guideline GRI 4 is clustered into criteria for reporting, reporting principles and reporting contents (standard disclosures). The standard should be implemented with the principle of disclosing or explaining, whereas the definition of explaining is, explain, why the company does not report according the GRI 4 standard (GRI, 2013, p. 2, 13). GRI reporting

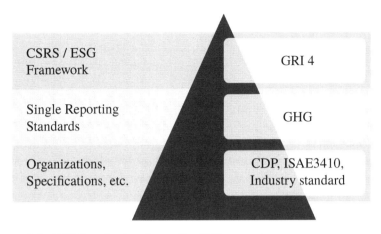

Figure 2.3: GRI Standard as frame for CSR reporting.

can distinguish between two different levels of detail. Core information is the information all companies need to report to be able to state that they are reporting in accordance with GRI 4. Comprehensive information must be published with regard to the material issues within the industry, so that the company is better able to meet stakeholder information needs. To explain how information shall be disclosed, the GRI defines principles for report content and principles for report quality (GRI, 2013, p. 11-15).

The CONTENT-BASED PRINCIPLES are stakeholder inclusiveness, sustainability context, materiality, and completeness. These can be summarized, that a company has to listen to its stakeholders as to what information to disclose, put this information into the broader context of CSR and identify the material issues based on the impact on economic, environmental and social impacts in accordance with the assessment of the stakeholders. The company has to ensure that the boundaries of reporting are clear and that the reporting completely covers the issues within these boundaries (GRI, 2013, pp. 16-17).

The QUALITY-RELATED PRINCIPLES contain a balance between negative and positive aspects, must be comparable over time and across organizations, to assess the performance in an accurate and detailed way, provide the information in timely fashion, be understandable and accessible to stakeholders, and provide reliable information which can be assured externally (GRI, 2013, pp. 17-18).

The GRI section called GENERAL aims to show the extent to which CSR is embedded in the company. This general section is divided into seven parts.

The first part is about the strategy and analysis which can be ensured with a letter of the Chief Executive Officer (CEO) describing how CSR is embedded in the strategy, and how this is operationalized. It also refers to the stakeholder environment of the company and their requirements. The scope of this letter should be the current state and a future outlook. This is followed by the second part, the organizational profile. This implies master data and more critical questions on the company, as the company needs to report on diversity KPIs, its relationship to union agreements and commitments to external initiatives. Questions on participation in governmental issues, strategic memberships, etc., also have to be disclosed. The third part reviews the identified material aspects and boundaries of the report with regard to the explicit question as to how these differ to financial boundaries of the company and how the boundaries have changed over time. Afterwards, stakeholder engagement shows how the company engages with stakeholder groups in preparation for identifying issues for this report. The report profile gives an overview of the period covered by the report, as well as the GRI indices covered and still outstanding. It also gives an overview as to whether the report has passed external assurance and, if so, with which level of assurance. From there the governance describes how CSR reporting is governed within the company, who holds responsibility for which parts, and who is incentivized on CSR topics. The last portion of the general part is about ethics and integrity. Here questions are asked about the code of conduct, interaction with external advisory bodies and the enabling of whistle-blowing or other escalation in case of unethical behavior (GRI, 2013, pp. 24-42).

After this general part, the GRI report consists out of different categories which are split into further aspects. For all of the substantive parts, the GRI starts with a so called Disclosure of Management Assessment (DMA) self-assessment. This DMA should explain to the reader why the respective content elements / indicators (e.g. carbon emissions) are material to the company and how the company takes this materiality into its management approach (GRI, 2013, pp. 45-46). The ENVIRONMENTAL CATEGORY is split into eight aspects as Figure 2.4 shows. These refer to energy, water, biodiversity, emissions, effluents and waste, products and services, (environmental) compliance, transport and overall information. Due to the focus on CO_2 disclosure, the emissions section is particularly interesting for the purpose of this research. The emissions are broken down by scope 1, scope 2, and scope 3. In addition, information regarding the GHG intensity ratio is asked, and about the reduction of GHG emissions (GRI, 2013, pp. 52-63). The emissions section is well aligned with the reporting requirement by the CDP to avoid

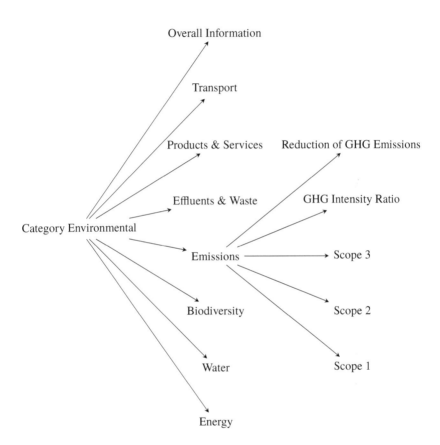

Figure 2.4: GRI 4 Category Environmental aspects.

double-reporting requirements (GRI, 2013, pp. 84-140). Overall, environmental information includes expenses for environmental protection, waste disposal costs, and assessments of supplier and environmental grievance mechanisms (GRI, 2013, pp. 52-63).

Whereas the GRI is a general framework on how to report CSR relevant information, the IIRC focuses on how to include non financial information in the financial statement. *The IIRC's mission is to establish integrated reporting and thinking within mainstream business practice as the norm in the public and private sectors. The IIRC's vision is to align capital allocation and corporate behavior to wider goals of financial stability and sustainable development through the cycle of integrated reporting and thinking (IIR, 2015).*

To achieve this, the IIRC is a partner organization of CDP, GRI, IFRS and IFAC (IIR, 2015). Figure 2.5 shows the structure of the IIRC. The IIRC council is dominated by accounting organizations and audit companies / persons with respective professional experience enriched with NGO, business background and executives from partner organizations. The core outcome of the IIRC is the IIRC FRAMEWORK. This shall help to establish more integrated thinking and reporting within companies.

The primary purpose of an integrated report is to explain to providers of financial capital how an organization creates value over time (The, 2013b, p. 4). This statement of the IIRC framework includes two main pieces of information about the understanding of the IR framework. First, the intended audience of the framework are providers of financial capital. Second, the generation of value is not limited to financial value. Value is seen here as value for the company and value for all stakeholder groups. As relationship

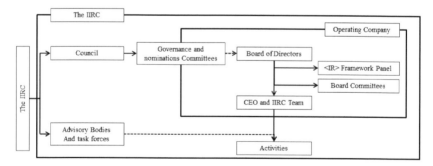

Figure 2.5: Structure of the IIRC (IIR, 2015).

management and other terms are hard to quantify into financial information, the IR framework uses the term of capital. Capital can be represented in several different KPIs, whereas financial capital is only one term among many.

Unlike the CDP and GRI, the IR framework does not list specific elements / KPIs which shall be reported but defines principles on how financial reporting can adopt non-financial KPIs. The focus of the framework lies on the value creation process of the company and how inputs are changed to outcomes through business activities, and how these activities are embedded in the business model (The, 2013b, pp. 4-13). As Figure 2.6 shows, the business model can input different capitals and output different capitals. The target of the framework is to show the respective interdependencies. Examples are that labour might become skilled during the working time, and nature can be destroyed or built up (e.g. in a company which plants trees), and goods can be transformed over time. As the IIRC is a partner organisation of GRI and IFRS, the general accounting principles apply, and the content of the report should be developed in a multi stakeholder approach. So the new core approach of this framework is to include, alongside financial capital, manufactured, intellectual, human, social/relationship, and natural capital in the decision-making process through integrated thinking (The, 2013b, p. 33). There are critical voices on the IIRC approach due to the strong investor focus, so that the integrated thinking is actually not going to broaden thinking about environmental issues but rather includes financial risks/opportunities that relate to environmental or other societal issues in the decision-making processes, which can be seen as the inclusion of a balanced-scorecard approach (Kaplan and Norton, 1992) in the financial reporting activities (Brown and Dillard, 2014, pp. 1120-1156). This argumentation is supported by Stubbs & Higgens analysis, which did not find radical or transformative changes in the integrated reports to date (Stubbs and Higgins, 2014, pp. 1068-1089).

In the LEGAL ADOPTION OF THE IR FRAMEWORK, South Africa is the piloting country with the King II report which forces all companies listed on the Johannesburg stock exchange to publish an integrated report (IIR, 2014b; Roberts, 2015, pp. 1-2). In Germany, there are several studies on how to integrate non-financial information into the general accounting process (Schmidt, 2012a). Where integrated reporting includes information useful for decision-making it is to be reported in the annual report. Two dissertations in 2012 showed that this is technically possible for annual reports (Schmidt, 2012b; Stawinoga, 2012). The Schmalenbach-Gesellschaft has already set up a working group to review options on how to do integrate

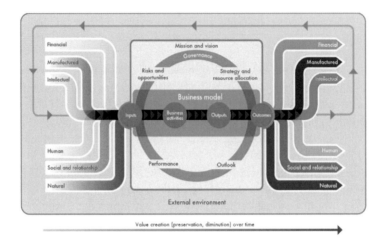

Figure 2.6: Value creation process (IIR, 2014a).

reporting in Germany (Arbeitskreis Externe Unternehmensrechnung der Schmalenbach-Gesellschaft für Betriebswirtschaft, 2015), and *Deutsches Rechnungslegungs Standard 20 Konzerlagebericht* (DRS 20) has already requested inclusion of non-financial KPIs which are used for management purposes (Bundesministerium der Justiz, 2012).

B. Climate disclosure initiatives

The most prominent CO_2 emission reporting initiative is the Greenhouse Gas Protocol (GHGP). This initiative defines the reporting needs of companies along their valur chain / product life cycle. Figure 2.7 shows the different activities during which emissions can occur. The top area of the figure lists the six different greenhouse gases according Kyoto Protocol Annex A (Kyoto-Protokoll, 1997, p. 13). This paper defines CO_2 emissions as the sum of the six gases normalized to the climate impact of CO_2e. Depending of standards and reporting elements, either only CO_2 emissions or CO_2e emissions are reported. The quantified list of an organization's GHG emissions and sources is called the inventory (WBC, 2004, p. 99). The area below the clouds in the figure shows the activities during which the emissions are created. These activities are clustered into three main scopes as defined by the greenhouse protocol (WBC, 2004, p. 25):

Scope 1: Direct GHG emissions occur from sources owned or controlled by the company, for example, emissions from combustion in owned or controlled boilers, furnaces, vehicles, etc.; or emissions from chemical production in owned or controlled process equipment.

Scope 2 accounts for GHG emissions from the generation of purchased electricity consumed by the company. Purchased electricity is defined as electricity purchased or otherwise brought into the organizational boundary of the company. Scope 2 emissions physically occur at the facility where electricity is generated.

Scope 3 is an optional reporting category that allows for the treatment of all other indirect emissions. Scope 3 emissions are a consequence of the activities of the company but stem from sources not owned or controlled by the company. Some examples of scope 3 activities are extraction and production of purchased materials; transportation of purchased fuels; and use of sold products and services.

As Figure 2.7 shows, own emissions can occur locally at a facility or in transport activities in vehicles of a company. The scope 3 emissions can occur in two ways. Upstream emissions relate to all activities in the value creation process before the reporting company receives the respective goods, while downstream activities occur after the reporting company hands over the product or the service.

After defining the different scope and the respective activities of GHG emissions, reporting companies have to define their operational boundaries. This should answer the question on what activities the reporting company need to account for, into which scopes the respective activities shall be classified and how the organizational boundaries are influenced by the consolidation methods. Companies can choose between Equity share, Financial control and Operational control as defined in the GHGP (WBC, 2010).

Under the EQUITY SHARE *approach, a company accounts for GHG emissions from operations according to its share of equity in the operation. The equity share reflects economic interest, which is the extent of rights a company has to the risks and rewards flowing from an operation.*

Under the FINANCIAL CONTROL *approach, a company accounts for 100 percent of the GHG emissions over which it has financial control. It does not account for GHG emissions from operations in which it owns an interest but does not have financial control.*

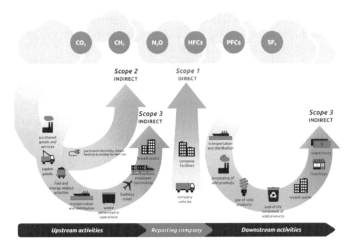

Figure 2.7: Overview of GHG Protocol scopes and emissions across the value chain (WBC, 2010).

Under the OPERATIONAL CONTROL *approach, a company accounts for 100 percent of the GHG emissions over which it has operational control. It does not account for GHG emissions from operations in which it owns an interest but does not have operational control (WBC, 2010).*

The effects on what emissions to account for, under which circumstances, can be found in the GHG Scope 3 reporting standard (WBC, 2010). For the further research, it is important to understand that the decision about the consolidation standard can have impacts on the classification of emissions between scopes 1 & 2 emissions and scope 3 emissions. After understanding the scope and the operational boundaries of the GHG inventory, it is important to understand which guiding principles are relevant for the accounting of GHG. The GHGP encourages companies to account for the target of a faithful, true and fair view. To enable this, the GHGP defines five REPORTING PRINCIPLES (WBC, 2004, p. 7).

RELEVANCE ensures that the GHG inventory appropriately reflects the GHG emissions of the company and serves the decision-making needs of users - both internal and external to the company.

COMPLETENESS describes that all GHG emission sources and activities within the chosen inventory boundary are reported. Specific exclusion has to be disclosed and justified.

CONSISTENCY aims for consistent use of methodologies to allow for meaningful comparisons of emissions over time. Changes in data, inventory boundary, methods, or any other relevant factors in the time series shall be documented transparently.

TRANSPARENCY addresses all relevant issues in a factual and coherent manner, based on a clear audit trail. Any relevant assumptions have to be addressed and appropriate references to the accounting and calculation methodologies and data sources used shall be made.

ACCURACY intents to ensure that the quantification of GHG emissions is systematically neither over nor under actual emissions, as far as can be judged, and that uncertainties are reduced as far as practicable. Sufficient accuracy to enable users to make decisions with reasonable assurance as to the integrity of the reported information shall be achieved.

These principles are very similar to the GRI reporting principles (GRI, 2013). However, the GHGP does not highlight timeliness, reliability or stakeholder communication (WBC, 2004, pp. 6-9). Applying the GHGP standard is typically used to account for emissions. The wording is used synonymously to carbon accounting, and as the GHGP does not provide a definition of carbon accounting within this research the following definition is used:

> Carbon accounting comprises the recognition, the non-monetary and monetary evaluation and the monitoring of greenhouse gas emissions on all levels of the value chain and the recognition, evaluation and monitoring of the effects of these emissions on the carbon cycle of ecosystems (Stechemesser and Guenther, 2012).

As the GHG Protocol is only a reporting framework, there are several guidelines and standards used to specify how the emission accounting shall take place. These standards are meant to reduce the amount of companies' judgement and make reports more comparable between companies (Auvinen et al., 2011b; Lewis et al., 2014; TNO et al., 2012, e.g.).

After companies assess their carbon inventory according to GHGP, they might want to disclose this information externally. One common way to publish the CO_2 inventory is to fill out the CDP QUESTIONNAIRE.

The CDP, formerly Carbon Disclosure Project, is a non-profit NGO. Its target is to help companies and their stakeholders measure, manage and share the companies' environmental information. It is supported by over 800 institutional investors managing a total of 95 trillion dollars (as of December 2015). With this institutionalized investor power, CDP aims to make the way companies do business more climate-friendly. One success is that the CDP database is the largest collection of self-reported data regarding climate change. This global data base enables investors and other stakeholders to take decisions based on environmental data while giving regulators data to enable policy changes (CDProadmap, 2015, p. 3, GRI2015, p. 2). The database is heavily used for climate-relevant research. As of November 2015, a literature review showed that over 150 research papers in 70 peer reviewed papers used CDP data (CDP, 2015a).

To enable stakeholders to judge on the reported data the CDP publishes the disclosure score and the performance score, and, as of 2016, the performance band calculated by the CDP. These are calculated based on information self-reported to the CDP, which also publishes the grading scheme allowing companies to self-asses their CDP score upfront. To understand if the data in the CDP are valid to use for taking environmental investment decisions, it is important to understand what information the CDP collects (CDP, 2015h, pp. 2-4).

In GENERAL, the CDP asks companies to use the GHGP for guidance purposes (CDP, 2015i, p. 1). The CDP questionnaire is structured into four main blocks. These are Management, Risk and Opportunities, the emission data itself and the approval of the submitted data. MANAGEMENT questions collect information about the highest level of management involved, incorporation into the incentive scheme, and how climate risk management is incorporated into the risk management of the group. It also examines how climate change is included in the business strategy. Additionally, it asks how the company interacts with regulators and trade associations in respect to climate change. In order to prove the content of the strategy, it asks about the targets a company has given itself to reduce carbon emissions and how carbon information is published externally.

The RISK AND OPPORTUNITIES block questions whether companies have identified risks or opportunities that generate a substantive change for the business operations, expenditure or revenue (CDP, 2015e, p. 9). All risks and opportunities identified need to be specified further in terms of drivers for the risk, likeliness and elements concerning the impact and the avoidance / realization of investments.

The EMISSION DATA block is the core part of the questionnaire. It starts with the base year and the base year emissions for scopes 1 & 2, as most climate targets relate to a certain base year. The questions refer to the methods used to quantify CO_2 emissions, in the CDP terminology global warming potential, and the respective operational boundary chosen. As soon as this is answered, companies can answer the question about the total scopes 1 & 2 emissions caused in the respective year and if these exclude any emission sources within the respective boundary. The next element asks for a self-assessment of data accuracy and information as to whether and how the data are audited. Afterwards, scopes 1 & 2 emissions are further broken down by energy type and geographic/legal distribution. The changes in emissions compared to previous year needs to be explained with reasons (e.g. reduction in activity or changes in boundaries, method, etc.). Afterwards, the CDP asks for emission-intensity KPIs. These are compared to (full time employees) FTE, compared to total revenue and a KPI in which the company can enter its own emission-intensity KPI or corresponding sector KPI. The last question in the emission block asks about the usage of carbon trading initiatives.

The SIGN-OFF / APPROVAL of data submission block simply enquires as to the highest management level which signed / approved the data submission to the CDP (CDP, 2015f, pp. 4-17). As the CDP focuses on the aspect of carbon reporting (even though the scope is currently expanding to additional environmental data, the core part of information is still the carbon questionnaire), there are other complementary initiatives that help investors, analysts, companies, regulators, stock exchange and accounting companies bring climate change-relevant information into the general scope of investment decision.

Another so called super-star organisation (Andrew and Cortese, 2013, p. 405) to standardize environmental, and especially carbon disclosure, is the CDSB. The CDSB was founded based on the initiative by the WEF in 2007 with the aim to harmonize carbon disclosure (Fleming, 2007). According to the CDSB, this should enable investors to make informed judgement around companies' risks and opportunities resulting from climate change. Investors should use the framework to include environmental information in their financial reports, and auditors shall audit against the framework. Regulators and stock exchanges should adopt the framework into their disclosure requirements (CDS, 2017). The WEF wanted to leverage the largest existing database for carbon reports, the CDP (Andrew and Cortese, 2013, pp. 402, 405).

The CDSB is made up of nine board members coming from different institutions. The board is chaired by a representative of the WEF. The other board members come from following organizations. *The Climate Registry (TCR), which is a non-profit organization governed by U.S. states and Canadian provinces and territories. TCR designs and operates voluntary and compliance GHG reporting programs globally, and assists organizations in measuring, verifying and reporting the carbon in their operations in order to manage and reduce it. TCR also consults with governments nationally and internationally on all aspects of GHG measurement, reporting, and verification* (The, 2015a). The International Emissions Trading Association

(IETA), which is a non-profit business organization of major companies and the leading voice of this community on emissions trading, whose goal is to ensure that the objectives of the United Nations Convention on Climate Change and, ultimately, climate protection, are met. CERES, an investor NGO network caring about climate risks and opportunities. Climate group, a non-profit NGO with the aim to develop a low-carbon future and the CDP. The World Resource Institute (WRI), which together with the World Business Council for Sustainable Development (WBCSD) publishes the GHGP. The WBCSD is also a member of the IIRC working group. And one advisor representative from the CDSB technical working group, currently led by a person who holds a job at an audit company (CDS, 2015a). The board is supported by a technical working group and an advisory board. Both the working group and the advisory board are dominated by industry members and investor initiatives, as Figure 2.8 shows.

In 2015, the CDSB published a framework on how companies can report environmental information into general audited financial reports with the aim to help companies to disclose their natural capitals in a standardized way. It shall help companies publish information while at the same time helping them comply with different legislation requests. This is done by reviewing existing standards from the CDSB members and existing legislation on disclosure. The connection between NGO standard setter and regulators shows that voluntary initiatives on disclosure have the potential to change the legislative environment. The exact content of the Framework is not discussed in detail (CDS, 2015b).

C. Carbon legislation

I. General ways of regulation

In order to understand the legislation companies are facing, it is important to understand the different kinds of legislation. Emissions can be regulated in several ways. The CDP lists *Requirements on energy efficiency, Production and usage of green energy, Climate Finance, Resilience, Emission trading systems, CO_2 taxes, Required carbon emissions reporting* (CDP, 2014a, p. 38) as paths for legislation. Additionally, each regulation needs to specify which of the gases to include and to which industry sector, and geographic scope the regulation is applicable (Kennedy et al., 2015, pp. 17-19).

Examples for MINIMUM REQUIREMENTS on energy efficiency can relate to products or product groups, production methods, waste disposal, and logistics chains. For products or product groups, minimum standards on energy and emission efficiency are very frequent. E.g. light bulbs in Europe must attain a certain energy efficiency level. The same applies for cars in North America and Europe for GHG emissions (ECOpoint Inc., 2015; Kuehl, 2013). Production methods also need to meet minimum requirements. One example is the enforcement of ISO certifications in production sites. E.g. ISO 50001 and ISO 14001 (DIN Deutsches Institut für Normung, 2011, 2009; Koe and Nga, 2009) provide transparency and enforce targets for full production, and ISO 23045 assesses the energy efficiency of buildings (Teng et al., 2014). Alternative to implementation based on ISO standards, regulators can also force companies to adopt other methods of clean production. E.g. by requiring certain filters in production sites or limiting the input factors for energy usage (Bundesumweltamt, 2013). There are several ways to regulate waste disposal. Usually, the target is to reduce landfill or avoid any other negative environmental impacts, e.g. by ensuring that hazardous waste is handled and disposed of in a way that minimizes risks to the environment. Examples for waste regulation can be found on the governmental overview of waste legislations and regulations of the UK government (Department for Environment Food & Rural Affairs, 2014). Logistics chains are often regulated locally. This can be the city, which requires cars to have certain filter technology available, or the ban on fossil driven vehicles in certain areas within a city (The, 2013c). Other transportation modes are also regulated, e.g. sea freight (DPA, 2013).

The production and usage of green energy can regulated via bio fuel amendments as required in the European Union (Deutsche Energie Agentur, 2006), green energy targets, grants for the production of green energy (Die Deutsche Bundesregierung, 2015) or other supportive actions to help the producers or consumers of green energy.

Climate finance denotes all financing flows associated with GHG mitigation activities. From a regulatory view, it is interesting that the government can directly or indirectly support mitigation activities with public money or set incentives for companies to invest their own money or join public-private partnerships to spend climate finance money. Often, these climate finance options create a money flow across borders, especially from more developed countries to less developed countries (Buchner et al., 2011, pp. 1-2). Figure 2.9 summarizes the different stakeholder and financial flows which can be summarized under climate finance.

There is academic consensus that climate change will VERY LIKELY have a significant impact on the physical environment in which companies work around the globe (IPCC, 2014, pp. 12-17). The identified core risks caused by climate change are issues resulting from sea level rise, coastal flooding, inland flooding and extreme heat periods. These issues include, among many others, the breakdown of infrastructure networks and critical services, food & water insecurity, and loss of ecosystems and biodiversity. As especially poor countries are more severely affected than rich countries, an increase of migration is expected (IPCC, 2014, p. 25). To prepare society and companies for these future challenges, regulators can set incentives for companies or invest directly in the respective measures (Guenther, 2009, pp. 70-74). One measure to reduce carbon emissions cost effectively is to stimulate reductions based on market behavior. The core ways to stimulate these reactions is to put a price on carbon emissions. This can be done either by setting a tax on carbon emissions or implementing an emission trading system (ETS).

II.   Emission trading systems and CO_2 taxes

ETS set a cap on the total level of emissions. Companies have to purchase emission polluting rights via an auction over time while the total number of ETS is limited. The first distribution of emission allowances can be done either via free allowances or direct purchasing models or a combination of both. Because emission allowances must be reduced over time, companies must reach certain abatements or purchase emission allowances from other companies. The marginal abatement costs to reach the total reduction target should lead to the market price for carbon. The core advantage of this system is that the reduction level can be fixed from a regulatory perspective while the market has the potential to solve the reduction challenge on the lowest marginal price. The disadvantage of such a trading system is the uncertainty of the respective carbon prices (EU ETS showed high volatility compared to other commodities). Complexity in installing such a trading system and the requirement of having sufficient participants in the market limit implementation. The complexity results from the quantification of the covered GHG and the distribution to the current market actors over the timeline. Prevention of market failure (e.g. due to overallocation of emission allowances) can also increase complexity (South Africa National Treasury, 2013, pp. 9-13). Alternative to ETS are direct taxes on carbon emissions.

CARBON TAXES have two core benefits. One is that the government is able to gain tax revenue. Economists in the US therefore argue for a carbon tax and a concomitant decrease in income tax (Auerbach and Cutler, 2015). The other benefit is that carbon taxes ensure a stable price (Kennedy et al., 2015, p. 10). The stable price is important to reduce uncertainties around energy-efficient investments. The core disadvantage of carbon taxes is that there is uncertainty about the total emission reduction (Kennedy et al., 2015, p. 10). Legislation is currently evolving towards putting a price on carbon. Figure 2.10 shows the global coverage of countries putting a price on carbon by installing cap-and-trade systems or taxing carbon emissions. All carbon pricing programs require transparency of emissions in order to calculate the respective price tag for the companies. Therefore, one prerequisite is to install a fitting system of carbon emission reporting.

As the regulation context is changing dramatically over time, this chapter has only the aim to provide an understanding that there are regulatory bodies which force companies to disclose their GHG emissions. All countries which have carbon pricing policies in place require companies to disclose their respective GHG emissions, at least to governmental bodies. In case a regulator decides to implement mandatory emission reporting (e.g. as part of a carbon tax or carbon cap-and-trade scheme), the regulator has to set the boundaries. These consist, first, of the scope of regulation as to which gases are covered, e.g. CO_2 vs. all GHG vs. different scoping of gases. Additionally, there must be a definition as to the sectors to which the regulation is to apply. Looking at the whole value chain, the definition of the scoping and decision on how and if to treat upstream and downstream emissions has to be clarified. Especially the ignorance of upstream emissions can lead companies to transfer energy-intensive activities to geographic locations outside the trading scheme. These design features influence the complexities companies face in adopting the regulation. In addition, several governmental institutions force companies to disclose their emissions to the public. As of 2011, KPMG counted 142 country requirements within 30 countries related to sustainability reporting (Andrew and Cortese, 2013, p. 401).

The King III Report in South Afrika is the first mandatory requirement for an integrated report worldwide with the scope of companies listed at the Johannesburg stock exchange (JSE) (King, 2009, pp. 10, 50). In case companies do not follow any principle of the King III principles, they can opt to explain which principles have not been applied and provide reasons therefore (The, 2014, p. 2). This should also imply the information on how environmental aspects are implemented in the decision-making action and how environmental issues are incorporated into the companies' strategy. As sustainability KPIs are not completely standardized, companies are obliged to explain the KPIs and their implications and potential benchmarks. GRI guidelines are considered as excellent reference. As it is an integrated report, the financial principles for reporting are applicable. External verification/assurance is named as mandatory, even though it is more complex than financial assurance (King, 2009, pp. 92-93). PwC summarizes the requirements as shown in Table 2.1. The most famous example for integrated reporting items might be the EU DIRECTIVE ON DISCLOSURE OF NON-FINANCIAL AND DIVERSITY INFORMATION by large corporations of public interest with at least 500 employees, which was approved in September 2014 and must be enforced by 2017 in the EU Member States (CDP, 2015b, p. 57). The EU guidance states that these companies shall report *non-financial information in a special statement containing information on environmental matters, social- and employee matters, human rights issues, and anti-corruption and bribery matters* (European Parliament, 2014, pt. I 6). Under environmental issues, GHG emissions are explicitly named as an element which shall be reported, however it is not defined under which standard the GHG emissions shall be reported.

D. Audit standard for Greenhouse gas inventory

As with financial information, the credibility of a disclosure increases if an external party verifies its correctness. For carbon trading schemes, a third-party verification is usually mandatory. In addition, the audit process helps companies understand their risks and ensures completeness (CDP et al., 2011, pp. 5-7) (Kennedy et al., 2015, p. 4). Auditing GHG emissions requires a choice of the standard under which information on emissions should be collected (e.g. ISO 14000:2006 or GHGP) and definition of the audit scope similar to the key design features. Also, the audit standard (e.g. AA1000AS or ISAE 3000) and the assurance level (limited vs. reasonable assurance) has to be named.

The Kings II report identifies two main audit standards to audit environmental information. These are AccountAbility's AA 1000 Assurance Standard (AA1000AS) and the International Accounting and Auditing Standard Board's International Standard on Assurance Engagements (ISAE 3000). Both standards can be used together, as AA1000AS is audit process-focused while ISAE 3000 focusses on the results in terms of complete and accurate data (King, 2009, pp. 93). The AA1000AS standard is a sustainability-audit

Principle	Summary Recommendation
The board should ensure the integrity of the company's integrated report	A company should have controls to enable it to verify and safeguard the integrity of its integrated report. The board should delegate to the audit committee to evaluate sustainability disclosures. The IR should be prepared every year. The IR should convey adequate information regarding the company's financial and sustainability performance. The IR should focus substance over form.
Sustainability reporting and disclosure should be integrated with the financial reporting	The board should include commentary on the company's financial results. The board must disclose if the company is a going concern. The IR should describe how the company has made its money. The board should ensure that the positive and negative impacts of the company's operations and plans to improve the positives and eradicate or ameliorate the negatives in the financial year ahead are conveyed in the IR.
Sustainability reporting and disclosure should be independently assured	General oversight and reporting of sustainability should be delegated by the board to the audit committee. The audit committee should assist the board by reviewing the IR to ensure that the information contained in it is reliable and that it does not contradict the financial aspects of the report. The audit committee should oversee the provision of assurance over sustainability issues.

Table 2.1: Recommendations by King III report for Transparency & Accountability (PriceWatershouseCoopers, 2009, p. 62).

standard which was written in a multi-stakeholder process including standard setter, audit companies and NGOs. It specifies the information the audit engagement shall include, what activities the auditors shall perform and how the scope of the assurance engagement shall be documented. This also includes subsequent publication of the respective information and the audience expected. In substantive terms, it specifies that the audit engagement shall define the level of assurance and if the assurance shall only test the principles of the information shown or also audit that the data/information is accurate and correct (Acc, 2008, pp. 5-16).

ISAE 3000 is a principle-based standard which defines principles, formats and information needs for assurance statements for sustainability reports. It specifies how the audit team shall be skilled, e.g. by multi-disciplinary teams including subject matter experts (IAA, 2011, p. 51). The assurance statement shall include a description of *any significant, inherent limitations associated with the measurement or evaluation of the underlying subject matter against the criteria* (IAA, 2011, p. 52) and highlight to the reader what the purpose of the assurance is and indicate if the subject-matter information is not suitable for other purposes. ISAE 3000 defines two levels of assurance: reasonable assurance and limited assurance (IAA, 2011, pp. 4-12, 52).

As result of the REASONABLE ASSURANCE engagement the auditor confirms the practitioner's information with an acceptable low level of remaining uncertainty (IAA, 2011, p. 19). For this reason, the auditor shall identify material risks, review the controls, test selectively, and potentially check with external information if the subject-matter information is correct. However, the aim is not to reduce the level of uncertainty to zero (IAA, 2011, pp. 28, 39). In contrast to that, the minimum requirement of a LIMITED ASSURANCE is that the information published should be meaningful to the set of data users (IAA, 2011, p. 4-12). Compared to reasonable assurance, limited assurance has a greater risk of incorrect information. Therefore, the assurance statement confirms that *nothing has come to the practitioner's attention to cause the practitioner to believe the subject matter information is materially misstated* (IAA, 2011, p. 28-29). This 'nothing has come to' has to be seen in the context that the auditor shall analyse in what areas material misstatements could occur and review them to an extent that the information can be used for the recipient of the information. As ISAE 3000 uses the term subject matter information the assurance on GHG emission was further specified in ISAE 3410 Assurance engagements on GHG statements.

The ISAE 3410 standard is the subject matter audit standard for the audit of GHG emission information. It does not cover potential efficiency KPIs, baseline information of life cycle assessments or instruments for emission reduction (ISA, 2011, p. 5). The auditor needs to comply with the requirements of ISAE 3000. ISAE 3410 is a specification for the element of GHG emissions based on ISAE 3000. As the GHGP, the ISAE 3410 asked if the boundaries are acceptable. The auditor has to check whether the methods for calculating GHG inventory are appropriate. Consistent with the CDP questionnaire, the audit standard checks for material changes due to mergers and akquisitions and divestment activities. The ISAE 3410 also differentiates between limited and reasonable assurance. The difference focusses on the entity's risk-assessment process and the implemented controls of the entity (ISA, 2011, secs. A82 & A84). For reasonable assurance, the critical assessment of the control environment is specified in greater detail. The testing of controls is only relevant for reasonable assurance. If a company estimates emissions, the auditor has to evaluate whether the methods are appropriate and test the method, check assumptions for reasonability, and self-develop a range of acceptable outcomes (ISA, 2011, pp. 9-14).

As GHG emission accounting involves scientific uncertainties and estimation/measurement uncertainties, the auditors expect a gap between the disclosed GHG emission and the real emission number. ISAE 3410 refers to this difference as the 'misstatement'. The task of the auditor is to ensure that the cumulated misstatement is not material. The question of materiality is not quantified but rather should be self-assessed by the auditor. The criteria are whether decision-making might be influenced due to the aggregated misstatement, and the misstatement shall be judged based on surrounding circumstances. Additionally, it is has to be answered if the accuracy of the emission information is reasonable for the intended user group or the majority of users. As user interests might differ, single users are not seen as the intended user group. User groups are defined as investors, market participants, regulators and management with their common interests in the company. As the method of publication also indicates the user group, e.g. the annual report focusses on investors, the intended method of publication provides the auditor an idea of intended user groups and their information needs. The auditor is allowed to focus on the major stakeholders with common interest. Depending on GHG elements and scope, the emission type, the nature of disclosure, the presentation method and the volatility of the emissions, the auditor can set thresholds and minimum requirements for materiality (ISA, 2011, pp. 9, 37-41).

E. Relevant ratings for Carbon emissions used in research

To enable investors to make decisions based on CSR and/or environmental activities of companies, there are several rankings that help investors identify the respective companies. The background for such ranking and ratings is the increasing demand from investors to invest in environmentally sustainable companies (The Carbon Disclosure Project, 2015, pp. 9-22). This chapter describes the indices often used in qualitative research to measure ESG and CSR outcome and aiming to inform investors. The Dow Jones Sustainability Index (DJSI), the MSCI ESG, formerly KLD 400 Index and the FTSE4Good indices are CSR rankings, while the CDP environmental leadership index is an example of a purely environmental index.

The DJSI is one of the most famous sustainability indices. It was launched in 1999 globally. Further sub-indices have been developed over time. The main target of the index is to track the performance of leaders in sustainability over time. The DJSI picks best-in-class companies in the field of sustainability (Searcy and Elkhawas, 2012, pp. 81-82). This positive screening is based on a standardized questionnaire that companies fill out and which is afterwards reviewed by the service provider RobecoSAM. The weighting, scoring criteria and best-in-class responses are not publicly available (Fowler and Hope, 2007, pp. 245-250).

The MSCI ESG and Domini 400 Social, index is set up for investors who want to exclude companies operating in sectors seen critical by certain investor groups. These sectors are nuclear power, weapons, gambling, etc. The negative screening is combined with a positive screening of companies with good ESG performance. To make the index comparable to the peer index, the sector and company size representation of the core index is applied. The index is reviewed on a quarterly basis. Launched in 1990, the index is one of the oldest ESG indices (MSC, 2015a, pp. 1-2).

The FTSE4Good index is also developed in a two-step approach. First, companies which produce weapons or tobacco are excluded; then, each company is assigned an ESG rating. If it falls below a certain rating, the company is delisted; above a certain rating, the company is listed (FTSE, 2015a, pp. 2-7). The rating criteria are set up based on the FTSE ESG advisory committee. Stocks are weighted to ensure liquid market conditions (FTSE, 2015b). The index is updated twice a year. As the other indices, FTSE4Good calculates indices for several markets, for example the US, Europe, UK, JP or emerging markets (FTSE, 2015a, pp. 3, 11-13).

Based on the CDP questionnaire, the CDP calculates three ratings. These are the Carbon Disclosure Leadership Index calculated on the carbon disclosure score, the Carbon Performance Index, and from 2016 these scores are merged to a performance band. The scoring of each index is documented in a scoring methodology document. This document describes how many points can be retrieved for each question, and examples are shown of ways to garner the maximum score (The, 2015b, p. 7). The carbon disclosure leadership index checks whether companies have implemented an effective data management system for energy consumption and GHG emissions. According to Paul Simpson, CEO of the CDP, this enables companies to spot their impact on climate change and see their risk and opportunity profile relative to climate change (Eckmueller, 2013). Ten percent of the companies within their benchmark companies are displayed as the leadership group. The carbon performance index indicates how successful companies implement their GHG reduction strategy. From 2015, there will be no longer a leadership group but the ranking will be placed in grades. Grade / band A is the best rating. One prerequisite of scoring maximum points is to score maximum credits on the development of absolute scope 1 & 2 emission, maximum points on the assurance of these emissions and publication of answers to the CDP questionnaire (The, 2015b, p. 5).

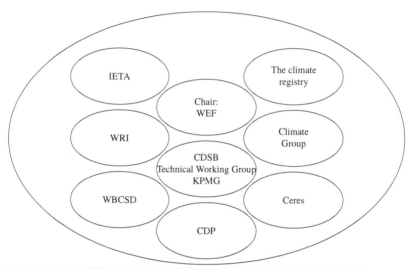

CDSB Advisory Committee & Technical Working Group					
Audit	**Companies & Investors**	**Standard Setter**	**Independent & Research**	**NGO**	**Government Bodies**
14 Members	8 Members	6 Members	6 Members	4 Members	3 Members
7 Big four 7 Professional network	CDP as investor Network 5 consulting companies Nestle & Maersk				

Figure 2.8: Structure and make-up of the Climate Disclosures Standards Board (Andrew and Cortese, 2013, p. 403).

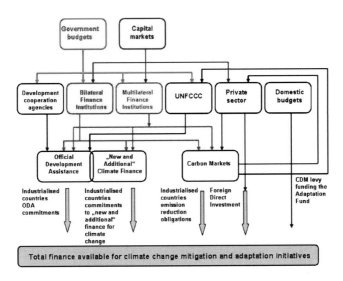

Figure 2.9: Financial flows for climate change mitigation and adaptation in developing countries (Atteridge et al., 2009, p. 4).

Figure 2.10: Carbon pricing programs around the world (Kennedy et al., 2015, p. 17).

Chapter 3: Relevant aspects of the carbon accounting literature

A. *Outcome from meta analytical studies*

Research mainly focuses on the connection between ESG reporting and financial benefits. One dominating research field is the effect of Carbon Accounting. Typical research questions in this field contain the question, if certain green actions pay off. This literature stream has several sub-categories that look at outputs such as the effect on reputation, costs, revenue, etc., or how influence factors for disclosure, such as company size, location, regulation, management structure, etc., affect the combination of financial and ESG performance. Less often, there is a review of the extent to which a disclosure relates with positive ESG changes on the part of a company. In addition, there are also several publications on what information is published, the impact of ESG criteria or rankings have, and the methods for information disclosure. There are also research streams that examine standards and regulations and their impact on companies. The hot topic here is currently the Integrated Reporting debate and the extent to which this changes the reporting of (non-)financial disclosure. Figure 3.1 based on the input of several literature reviews (Ascui, 2014; Fifka, 2013; Gray et al., 1995; Hahn et al., 2015; Hartmann et al., 2013; Healy and Palepu, 2001; Podsakoff et al., 2003; Rom and Rohde, 2007; Stechemesser and Guenther, 2012; Zvezdov and Schaltegger, 2014)and meta-studies (Albertini, 2013; Dixon-Fowler et al., 2013; Endrikat et al., 2014; Fifka, 2013) shows the current research areas. As shown in Figure 3.1 the research object of GHG emissions is reviewed for output factors such as what information is disclosed, how it is disclosed, what standards are used by which stakeholders and so forth.

Another research stream is around questions of the determinants that prompt companies to disclose carbon information. This is often done based on quantitative studies reviewing the relationship between company characteristics and disclosure, less often in exploratory or qualitative settings to analyze the decision process of companies to disclose the respective information. Most studies, however, focus on the outcome dimension in terms of the impact of carbon disclosure on either the financial situation of a company or the impact on the carbon performance of a company which discloses their footprint. In terms of the theoretical framework in which carbon accounting

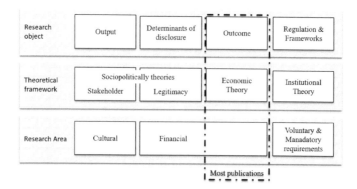

Figure 3.1: Research streams around Carbon Accounting (Gray et al., 1995; Hahn et al., 2015; Hartmann et al., 2013).

studies are embedded, there are several different viewpoints. Most critical from the literature reviews is the fact that there are several publications without any theoretical framework. In the studies including a theoretical viewpoint, stakeholder, legitimacy, economic theory (e.g. agency) and institutional theories are frequently used (Gray et al., 1995; Hahn et al., 2015; Hartmann et al., 2013). The studies focusing on the scientific error of carbon emissions are limited and often have an engineering background (Auvinen et al., 2011a). There are also a few works that critically review the disclosed data with regard to data quality (Green, 2012, Hartmann, 2013, pp. 558-559) or critically examine the ways in which accounting literature focuses only on technical elements but not on the way stakeholders interact to set up regulatory requirements (Andrew and Cortese, 2011a,b, 2013).

B.   The pays-to-be-green business case

I.   Argumentation for and against the green measures

If the mission of companies is governed by the principle that *Business of Business is Business* (Friedman, 1962), there seem to be limited reasons to perform green actions that use resources and are therefore not beneficial in profit terms (Friedman, 1970; Greer and Bruno, 1996; Jaffe and Peterson, 1995; Verhoef et al., 1996; Walley and Whitehead, 1994) as environmental costs are transferred to the company (Dixon-Fowler, 2013, p. 355, Bragdon,

1972, pp. 9-18). This research first provides an overview of the relevant stakeholders namely investors, regulators and environmental interest groups as target audience of environmental reports and maps these stakeholder interests and ways of influence to arguments. Based on these arguments, theories that fit the respective arguments are identified. This is in line with the interpretation of stakeholder theory as a framework that can be used to apply/develop additional theories (Parmar et al., 2010, p. 6).

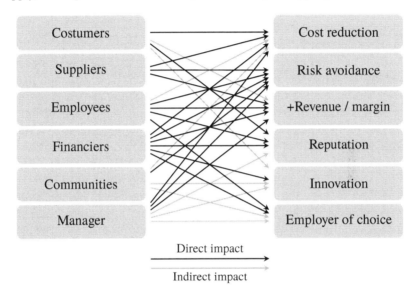

Figure 3.2: Influence of stakeholders on Pays to be green elements.

Figure 3.2 shows on the left side the various stakeholders a company has to deal with (Parmar et al., 2010, p. 5), and on the right side the business reasons (Schaltegger, 2013, p. 12) for green actions on the part of companies. The arrows show which stakeholder group benefits from which business reason. E.g. customers benefit if the companies have reduced costs, as there is a potential for decreasing prices or increasing profits, which benefits the investors. To limit complexity, interdependencies between stakeholders are not shown, and the stakeholder communities include the general public, e.g. NGOs, media and regulators. These stakeholders can benefit from a company's green actions. They thus have an interest in the company's performance of respective actions.

Cost reduction can be achieved by using less material/energy, thereby achieving a better cost base (Albertini, 2013; Botta et al., 2012; Dechant and Altman, 1994; Hart, 1995; Porter et al., 1995; Shrivastava, 1995). This benefits all stakeholders that depend on payments by the company, e.g. suppliers, employees, communities (higher taxes), customers (potentially lower prices). Especially stakeholders who participate directly in the success of the companies, e.g. financiers and managers, should ensure that the company improves its energy efficiency over time. One argument against the focus of reducing materials and energy use is that this focus leads to an over allocation of financial and management resources to minimize the energy/material costs while other cost categories (e.g. manual labor) are disregarded. This prioritization on energy efficiency could lead to higher costs overall (Hertwich et al., 1997). This relatively higher attention can be justified, as energy/material (and therefore waste) reduction leads to other positive side effects not reflected in the pure cost of material/energy/waste disposal (King, 2009).

Various stakeholders also benefit from the reduced environmental risk portfolio of the company. These risks can either result from legislative action (e.g. carbon pricing), legitimation risks (Bansal and Roth, 2000; Hahn and Luelfs, 2013; Hahn and Kuehnen, 2013; Hrasky, 2012; Milne and Patten, 2002a; Suchman, 1995), risks regarding the environment-friendly product portfolio (e.g. assets from coal mining companies decrease significantly in value in the case of a carbon tax). Especially the risk of increased legislative action is very likely after the Paris agreement 2015 (Kennedy et al., 2015; Kyo, 2014; Uni, 2015). This argument is still valid after the election of president Trump in the United States who supports actively the fossil energy industry (Greshko et al., 2017), as several ways of environmental regulation are done outside the US or on a different governmental level (e.g. city or state specific regulation) and many political bodies have a self interest to differentiate themselves from the Trump administration (Jacobo, 2017; Popvoich and Schlossberg, 2017). This argument is especially valid for the world wide biggest CO_2 emitting country China (Johnston, 2017). Therefore, all long-term stakeholders are interested in seeing the company reduce its environmental risk profile. If investors are concerned that a company is not managing environmental risks well enough, investors could start to divest their holdings in these companies. As an example, Allianz and Rockefeller decided to divest all companies from the oil and coal industry (Allianz S.E., 2015; Heaps, 2015; The, 2016). A reduced environmental risk portfolio thus reduces a company's cost of capital (Cheynel, 2013; Dhaliwal et al., 2009, 2011; Verrecchia, 1999).

Environmentally friendly actions can have a positive impact on the product portfolio, resulting in higher revenue and/or margins from which various stakeholders benefit. Customers can choose to buy products with less environmental impact. The assumption is that the old product portfolio could be still accessed (e.g. from competitors), and the customer has an additional value which results in the green premium. The effect of the extra premium can also turn negative if all competitors take respective green actions and the company is part of an adverse selection. As with cost reduction, all other stakeholders benefit from the respective extra rent. Measurement of how much more customers are willing pay for green products is limited, theoretically, however, the market share for environmentally friendly products can range up to forty percent (Balderjahn, 2013, pp. 13-20).

As the UN Climate Report points out, green actions can be used to increase brand value (UNG, 2015, p. 180). In general, this means that companies include their CSR activities in their communication strategy to increase brand value (Arnold, 2011, pp. 46, 185-200). This positive impact is also perceived by investors if there is a positive payback expected (Fiedler, 2007, pp. 282-283). A positive signal is that big companies identify reputation as one core driver to do CSR reports (KPMG et al., 2015, p. 18).

The employer-of-choice argument arises from the brand value / reputation argument. The idea is that employees are prouder of their companies if they know that their company takes their responsibility serious (Eccles and Saltzman, 2011, p. 60). The opposite effect is that potential employees refuse to apply for jobs with a company if the financial value creation process is not adding value for society/the environment (CDP, 2016a, p. 4).

Porter argues that CSR activities can increase innovation activities for companies where CSR activities are embedded in company strategy (Porter and Kramer, 2006; Porter et al., 1995). Especially the higher motivation of employees to do something environmental friendly and the new challenges are seen as drivers for innovation (Hockerts et al., 2009, pp. 44-47). Further arguments on how green actions increase innovation can be found in the literature as eco-innovation (Hockerts, 2009, p. 8, Hart, 1999, p. 82, Hawken, 2013, p. 1, McDonough, 1998, p. 82, Senge, 2001, p. 24). Precarious to these arguments is that green innovations are often the result of legislative pressure and that innovation only compensates for the higher challenges. In such cases financial performance should not be influenced (Aguilera-Caracuel and Ortiz-de Mandojana, 2013, p. 8).

Even though there are strong arguments that green investments pay off, there are also ARGUMENTS AGAINST GREEN INVESTMENTS. The product side

Pro	Contra
Cost reduction for energy, materials and waste	Management focus on green measures may lead to overallocation of resources
Reduced environmental risk e.g. by anticipating regulative actions	Transfer of environmental cost to the company
Increased eco innovation	Reduced potential product portfolio by shifting to green products
Better brand value for green products	Limited willingness from customer to pay a surplus for green products
Reduced cost of capital required due to reduced environmental risk portfolio and no adverse selection from investors	Green washing might be similar successful to the brand elements as real green measures
Increased employer branding	

Table 3.1: Arguments for and against green measures of a company

and the respective arguments are contradicted by the limited buying power of consumers purchasing primarly green products (Valor, 2008, p. 1). Additionally, the business case for green investments has a certain peak level at which it pays off (Mintzberg, 1983, p. 10). Therefore, some researchers argue that regulation is the key driver to force/convince companies to adopt green practices (Carroll and Shabana, 2010, pp. 100-101). The financial arguments for and against green measures from a company perspective are summarized in Table 3.1.

II.   Empirical evidence of the business case for green actions

As there are several arguments to why green actions will or will not pay-off, researchers are interested to understand whether investments in green technologies are profitable. To answer this question quantitatively, we must transpose green investments into elements that can be measured empirically. As this answer seems to be quite relevant in the environmental research community, there are plenty of studies that examine the extent to which CSR generally and environmental actions in particular affect a company's financial performance. Empirical evidence about the outcome of green investment is difficult to assess, however. The network for green business adoption of WE MEAN BUSINESS looks to the project level and arrives at an internal rate of return of 27 (Nicholls et al., 2015). But it is likely that this number is biased by the question and sample of projects, as companies are being interviewed to share their success stories (Law et al., 2011, p. 8).

Looking at the diverse set of stakeholder argument combinations, it explains why research uses different theories to explain green actions and/or voluntary environmental reporting. E.g. the approach of a research paper analyzing customers' willingness to pay an extra margin for green products (Hart and Dowell, 2011, pp. 1469-1473) might be completely different to that of a research question around a company manager in the chemistry industry who discloses the carbon footprint in the face of pending legislation (Milne and Patten, 2002a) or brand value (Hahn and Kuehnen, 2013; Haddock, 2006; Nikolaeva and Bicho, 2011).

In addition, there are numerous studies comparing the relationship between environmental/CSR actions and the financial performance of the respective companies (Clarkson et al., 2011; Fisher-Vanden and Thorburn, 2011; Fürst and Oberhofer, 2012; Greenwald, 2014; Haigh and Shapiro, 2011; Hart and Ahuja, 1996; Murguia and Lence, 2015; Russo and Fouts, 1997; Saka and Oshika, 2014). These studies are done either as event-based or long-term studies focusing on different regions, sectors, time-frames and measures for operationalizing CSR/green performance and corporate financial performance. Looking at the meta-analytical outcome of that research stream, there is usually a mixed but positive outcome (Albertini, 2013; Carroll and Shabana, 2010; Clark et al., 2014; Dixon-Fowler et al., 2013; Endrikat et al., 2014; Parguel et al., 2011). Clark, Feiner, and Viehs, for example, reviewed nine meta-studies and their underlying studies on the relationship between ESG & corporate financial disclosure and found a clear positive connection between the two. In addition, the relationship between environmental management and operational performance, and the relationship between environmental performance and stock price, are perceived as positive (Clark et al., 2014, pp. 29-38).

The Table 3.2 shows how many studies are reviewing the combination between ESG performance and the implications on the company. This very positive result might be of limited value, as there might be a conflict of interest between the publishing bodies, a consulting company with a focus on green investments, and the outcome. Other meta-studies often identify significant

Impact on	Impacted by	Number of studies	Thereof Positive
Cost of capital	Environmental	15	12
	Social	11	9
	Governmental	20	18
Operational performance	Environmental	29	23
	Social	27	22
	Governmental	27	23
Financial performance	Environmental	22	18
	Social	21	17
	Governmental	22	19

Table 3.2: Effect of ESG studies on corporate performance (Clark et al., 2014).

limitations in the pays-to-be-green research (Albertini, 2013; Dixon-Fowler et al., 2013; Endrikat et al., 2014; Hahn et al., 2015).

III. Critics on carbon accounting studies

To understand the limitations of meta-analytical studies, it is important to understand the dimensions of the relationship between CFP and CEP. One dimension considers ways of measuring CFP (e.g. stock price, EBIT, etc.) and environmental performance. This can be looked at from either a process or output perspective.

The Table 3.3 shows how green performance is operationalized in the ESG studies and in which journals these studies were published based on a literature review of Delmas et al. (2013). Here the dominace of the KLD index is visible. As the KLD score is influenced by environmental concerns relating to waste, regulatory problems, emissions and climate change and strengths such as products, pollution prevention, recycling, clean energy, communication and environmental management systems the usage of those scores might be heavily influenced by the sector and geographic scope of the production and distribution activities of the reporting company which makes an quantitative comparrison difficult.

Publication	Total	KLD	SAM	TRUECOST
Strategic Management Journal	10	10	0	0
Academy of Management Journal	11	11	0	0
Journal of Management	7	7	0	0
Int. Journal of Management	2	2	0	0
Business and Society	14	14	0	0
Journal of Business Ethics	31	28	2	1
Other	26	16	5	5
Total	101	88	7	6

Table 3.3: Operationalization of green performance (Delmas et al., 2013)

Sector	KLD 400 Social	USA IMI
Information Technology	27.5%	19%
Consumer Discretionary	13.93%	13%
Health Care	12.56%	14%
Financials	10.61%	17%
Industrials	9.9%	11%
Consumer Staples	9.09%	9%
Energy	4.99%	8%
Real Estate	4.41%	0%
Materials	2.81%	4%
Telecommunication Services	2.4%	2%
Utilities	1.79%	3%

Table 3.4: Sector weight of KLD- and reference index (MSC, 2017, 2015b)

Looking at the composition of the KLD index and the reference indices Table 3.4 it is obvious, that the index weight has a different sector composition as the underlying index. In other words, some studies do not measure how ESG influences the performance but rather how the US tech industry outperforms the US economy during a certain period of time. The results imply positive (Delmas et al., 2013), negative (Busch and Hoffmann, 2011) and mixed results in meta studies. Specifically, here Endrikat et al. (2014) and Albertini (2013) list toxic releases and pollution/prevention [Hart, 1996, pp. 29-38, Dixon-Fowler, 2013, King, 2002], compliance factors (Barth and McNichols, 1994), strategy incorporation (Sharma and Vredenburg, 1998) and environmental ratings (Wagner, 2010).

Table 3.5 shows an overview of what other elements are used to operationalize green measures (Albertini, 2013). These elements are often rather weak proxies with which to operationalize green performance.

Another crucial issue to measure the relationship between CFP and CEP is the kind of environmental action a company will take, as such actions could be done proactively or in response to stakeholder pressure. This question can affect the CFP-CEP relationship, as proactive strategies tend to have greater influence on a company's action. Additionally, the CFP-CEP studies usually do not account for timing differences between the green investment period and the financial gain. Due to the missing timing implications the causality for any CFP-CEP relation remains an open question as financial strength can be also associated with a higher ability to invest in innovation e.g. green actions [Dixon-Fowler, 2013, pp. 354-355, Endrikat, 2014, pp. 735, 741]. Another element refers to the often missing theoretical concept. This potentially leads to a bias if there is a circle of relationships between the CEP and CFP [Endrikat, 2014, p. 739, Hart1996]. Studies are often criticized for a lack of or insufficient moderator/control variables (Albertini, 2013, p. 431).

Aside from the proactive vs. reactive strategies, it is also recommended to account for company size, methodological issues regarding the source of the data (e.g. self-reported data at CDP), the headquarters location, ownership, the industry or the carbon intensity of the sector involved (Dixon-Fowler et al., 2013, pp. 354-356). This is because companies often face different stakeholder pressures based on the identity of their stakeholders, their views of the importance of environmentally friendly solutions and their stakeholder salience (Guenther et al., 2015). Even though the question of does it pay to be green might not be answered yes or no, the empirical evidence used to argue why it may pay-off to invest in green actions can be summarized to say that there are certain stakeholders which have a demand for green disclosure.

Environmental performance variable	Environmental management variables	Environmental disclosure variable
Total emissions	Integration of environmental issues into the strategic planning process (perceptual measures)	Environmental disclosure in annual report
Relative emissions	Environmental policy	10-K report
Industry emissions	EMS adoption	Environmental information in newspapers
Toxic Release Inventory (TRI)	EPAs voluntary 33/50 program	Environmental lawsuits
Carbon emission reporting	process-driven and product-driven environmental initiative	Ranking
End-of-pipe control pollution	environmental strategy (perceptual measure)	Environmental awards
Environmental reporting	ISO 14001 certification	
Pollution performance index	environmental practices	
Toxic chemical emission		
Pollution released		
Emission reduction based on TRI		
Waste recycling rate		
Waste reduction		
Toxic waste treated		
Greenhouse gas/ozone depleting substances emissions		
Emission of organic carbon		

Table 3.5: Proxies to measure environmental management (Albertini, 2013, p. 451)

C. Carbon disclosure as reporting element for green actions

Even though there are different environmental proxies, carbon figures and carbon accounting constitute a dominant area of research. The classification follows the framework of Hahn et al. (2015) as shown in Figure 3.3. The first section in figure describes the process used to disclose carbon emissions. This process is usually referred to as CARBON ACCOUNTING and covers the collection, preparation, measurement and processing of the respective data. The second area shows the output of disclosure and the respective determinants that influence disclosure. The third area concerns the impact of carbon disclosure (outcome), meaning what is influenced through the disclosure of carbon emissions. These sections are framed by the side conditions of carbon accounting concerning the regulations and initiatives under which carbon disclosure must occur.

The process towards carbon disclosure of the framework in Figure 3.3 is part of carbon accounting. One reason why there is not a single definition of CARBON ACCOUNTING is the existence of different frames caused by *different communities of practice: physical, political, market-enabling, social/environmental and financial carbon accounting* (Ascui, 2014, p. 8). These different communities can cause tensions in the interpretation of carbon accounting (Ascui, 2011, p. 982, Ascui, 2014, pp. 7-10). The definition by Schaltegger & Burrit is sufficient for purposes of this study. Here, carbon accounting is seen as (Schaltegger and Burritt, 2014, p. 234):

> activities, methods and systems [as well as] recording, analysis and reporting of environmentally induced financial impacts and ecological impacts of a defined economic system (e.g. company, plant, region, nation, etc.).

Each element of this definition is somehow a field of research unto itself, with research questions of its own. Studies with an activities based view usually cluster activities into a voluntary and a mandatory part that determine the output of the carbon accounting. In the research field of methods to assess a carbon inventory, questions arise regarding the complexity of the calculation and the robustness of the data. The inventory is limited by these models and assumptions. This influences the ability to audit the data and the respective systems (Ascui, 2014, pp. 7-10). The issue around recording the data is connected on the one hand to the scientific measurement error to count gases. On the other hand, similar questions to cost accounting in the accounting and accountability for carbon are discussed (Bowen and Wittneben, 2011, pp. 1026-1028). Analysis of carbon accounting can be done based on a

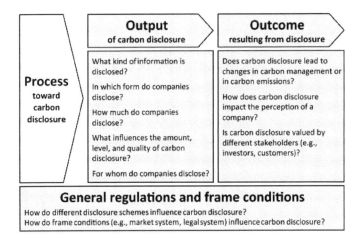

Figure 3.3: Process framework of carbon disclosure with exemplary research questions (Hahn et al., 2015, p. 82).

physical or monetary basis, past- or future-orientated, routinely generated or ad-hoc; short- or long-term outlook is possible as well (Ascui, 2014, pp. 7-10, Burritt, 2011). The analysis is often done based on reports which can be either internal - often described as carbon management accounting - or external. Internal reports are used to analyze the status of voluntary reduction commitments, cost accounting and respective resource allocation (Burritt et al., 2011). External reports are used to ensure compliance and benchmark companies, or inform stakeholders, especially regulators (Stein, 2009, pp. 291-310, Xuchao, 2010, pp. 4520-4527). In external reporting, the use of reporting frameworks and auditors is being researched (Stechemesser, 2012, p. 25, Olson, 2010).

The financial impacts measured by financial carbon accounting are also discussed from various angles. The general approach behind financial impacts of carbon emissions is to factor external effects into the profit and loss of the emitting company, also called financialisation of the atmosphere (McNicholas and Windsor, 2011, p. 1074). For companies, the question of carbon markets or carbon taxes is not as important as the question of how competitive risks and opportunities result from these. Thus, carbon management accounting must ensure that CO_2 intense processes are identified (and respectively reduced), and that low carbon-emitting activities and products are developed. Therefore CO_2 emissions must be included in business decisions, e.g. investment planning to derive a competitive edge from increased fictionalization trends (Ascui, 2014, pp. 7-10, Kolk, 2008).

In terms of ecological impacts, it is important to understand what emissions (which gases) shall be accounted for in which unit of measure. This is especially important if emissions are to be compared at a later point in time. The area of the defined economic system has a high impact on the research for the process for the disclosure. As named in the definition, the economic system can focus on various elements, e.g. companies, plant, region, nation, products, projects, etc. Depending on the economic system involved, it is likely that all other elements of the definition are different, as the target audience for a national carbon report is likely to differ from the audience for the carbon report of a factory/facility. The methods and data availability for emissions also differ based on the economic system. E.g. a coal-fired power plant has voluminous evidence about the amount of coal burned and the emissions generated, while the calculation of a national carbon footprint is done via a complex model that involves a number of assumptions. Looking at the frameworks for company-relevant disclosures, emissions trading schemes usually focus on the asset that causes the emission (e.g. plant), while reporting schemes as the GHGP focus on the whole company (Green, 2010).

Looking at the open questions resulting from the economic system definition (and thus affecting all other elements of the definition), the boundaries of an organization for carbon accounting need to be discussed further. One element here is the role of scope 3 emissions, which usually dominate overall numbers but are harder to measure and influence. Because they dominate totals, it is recommended to include the emissions in the decision-making process (Fallaha, 2009, pp. 885-893, Putt del Pino, 2006, pp. 1-69, Stechemesser, 2012, p. 33). Another research gap are field studies on how to connect the carbon inventory accounting and the carbon footprint accounting. A footprint calculation is currently defined as (Weidema et al., 2008, p. 4):

> Accounting for carbon footprints is a question of quantifying and presenting emissions data for the whole life cycle of products in a consistent manner.

This definition is consistent with the use in other research papers (Larsen, 2009, pp. 743-762, Stechemesser, 2012, p. 32-35).

The output perspective focuses on research to explain which determinants support carbon disclosure. Most studies follow a quantitative approach (Hahn et al., 2015). The identified determinants include economic, ecological and regulatory factors as well as other determinants of GHG disclosure. The results for carbon disclosure match mainly the results for ESG disclosure, e.g.

that large, financially successful companies disclose more information (Fifka, 2013). As an upcoming topic, voluntary assurance is connected particularly with the ISAE3410 standard. Additionally, options for managers to choose which indirect emissions (scopes 2 & 3) should be reported under what standard is an upcoming field in research (Hahn et al., 2015, pp. 89-91).

The outcome dimension can be split into the economic outcome of carbon reporting and the ecological outcome. The effect of carbon disclosure on economic performance and emissions performance is inconsistent but tends to be positive (Kim and Lyon, 2011). Critical to the application of the agency theory is that the effect depends on the regulatory environment. As most studies do not refer to any theoretical framework, this can be seen as an open research field (Hahn et al., 2015). In terms of ecological outcome, the results are also inconsistent in terms of trends in emissions and energy costs. This applies to carbon disclosure in general but also to the effort to join an energy reduction programme (Hahn et al., 2015, pp. 94-95). Accordingly, some prior studies see open research questions in the relationship of climate (action) strategy and climate disclosure (Bebbington and Larrinaga-González, 2008). However, once a company joins such a programme or starts to disclose information, that company is likely to continue to do so (Stanny and Ely, 2008, Stanny, 2013, pp. 338-348). This could be explained by increasing stakeholder interest as soon as the information is made available.

As carbon reporting consists of mandatory and/or voluntary elements, existing research also looks at different frames and at the extent to which they are dependent on one another. In general, results show that carbon reporting is more often voluntary (Hahn and Luelfs, 2013, pp. 401-417). Within voluntary reporting, the success of CDP & GHGP is often described as effective forces (Green, 2010, pp. 1-35, Knox-Hayes, 2011, pp. 91-99). Arguments against voluntary reporting are the limits of reliability, comparability and understandability of the information (Andrew and Cortese, 2011a), overestimation of the benefits for companies (Harmes, 2011; Sullivan and Gouldson, 2012) and hence the non-relevance for users (McFarland, 2009, p. 281).

As voluntary reporting fails to deliver perfect results, some researchers argue for more mandatory reporting (Erion, 2009, Raingold, 2010, p. 185). Increased mandatory reporting often increases levels of information provided on a voluntary basis as well. Mandatory reporting schemes often build upon existing voluntary reporting schemes (The, 2014). Therefore, some studies advocate combining mandatory and voluntary reporting schemes (Andrew and Cortese, 2011b; Knox-Hayes and Levy, 2011; Sullivan and Gouldson, 2012).

Alternatively, some authors argue that mandatory reporting should focus on climate risk disclosure. In this case, reporting needs are driven by financial reporting standards (Bebbington and Larrinaga-González, 2008; Burton, 2010; Doran and Quinn, 2009; Hahn et al., 2015; McFarland, 2009; Pattberg, 2012; Solomon et al., 2011). Studies within this research field are sometimes criticized for being only conceptual and having limited theoretical background or respective empirical studies (Hahn et al., 2015, pp. 87-89).

D. *Theoretical background of carbon accounting*

I. Stakeholder theory

Based on a recent literature review of carbon accounting literature, about half of the peer-reviewed journal articles in the field of carbon accounting do not refer to theories. Most of the remaining half uses one or more of the following theories (Hahn et al., 2015, pp. 87-89):

– Stakeholder theory,
– Legitimacy theory,
– Institutional theory,
– Agency theory.

Looking at the stakeholder theory, the analysis by Parmar et al. argued that stakeholder theory should not be interpreted as a single theory but rather as a framework grounded on the observation that companies operate within an environment together with other stakeholders. Stakeholder interests thus influence the way a company creates value (Parmar et al., 2010, p. 6). In terms of opportunity to influence the company, different stakeholders have different options and methods for doing this (Niehues, 2016, p. 7-11).

Generally, the stakeholder framework should be applied to explain the existence of CSR and the disclosure of CSR activities as the audience of voluntary disclosure are stakeholders (Parmar et al., 2010, p. 22). Disclosure of carbon information can thus be explained based on stakeholder interest in this information. The aim of the reporting company to disclose carbon information is to generate a positive view of the company (Cotter, 2012, pp. 169-184, Guenther, 2015, p. 5, Hahn, 2015, pp. 85-87, Hackston, 1997, pp. 77-108). Empirically, this explanation is supported by the finding that elements connected to stakeholder exposure, such as company size, media visibility and ownership structure, can explain the likelihood that a company will disclose information or not (CDP, 2014, p. 38, Searcy, 2012). Other supporting elements include the regulatory framework in which the company operates and the influence of activist groups. Profitability, capital structure and company size, are less suited to explain disclosure (Boesso and Kumar, 2007; Dienes et al., 2016; Wackernagel and Rees, 1997). Looking at pure disclosure, it is important for stakeholders for disclosure and performance to correlate (Clarkson, 2008, Guenther, 2015, p. 4, Guenther, 2014, p. 689, Huang, 2010, pp. 435-448, Prado-Lorenzo, 2010, pp. 391-420). As the stakeholder framework is often referred to as STAKEHOLDER THEORY, the interpretations explaining carbon disclosure can vary widely (Donaldson and Preston, 1995, p. 66). This is due to overlaps with other theories such as legitimacy theory in the event of social pressure (Guenther et al., 2015, 5-6), institutional theory in the development of the CDP (signatories and reporting companies) and homogeneous reporting information (Parmar et al., 2010, pp. 40-41), and agency theory in which the company sends positive signals to its stakeholders.

II. Legitimacy theory

The general assumption behind legitimacy theory is that an IMPLICIT SOCIAL CONTRACT (Patten, 1991) exists between society as a whole and the company. Therefore the company has to ensure that the social contract will remain in effect, and the company remains the licence to operate (Hart and Milstein, 2003). Conversely, if society is concerned about the business acumen of a company, this concern leads to pressure, and the company has to react. One option to react is to inform about social and environmental impacts of the company by disclosing the relevant information (Cho and Patten, 2007; Deegan et al., 2002; Gray et al., 1995; Ramanathan, 1976; D'Souza et al., 2004). If the social contract is breached, it is likely that society, which can also act dynamically, will take actions to penalize the company. This risk of penalizing power on the part of society ensures that the company will behave in compliance with the implicit social contract. If ESG disclosure is used to demonstrate that the company is compliant with the requirements of society, it is likely that this disclosure will communicate only positive information or use a positive verbal description for the required facts (Deegan and Rankin, 1997; Maines et al., 2002; Milne and Patten, 2002a,b; Patten, 1991; Shocker and Sethi, 1973).

As disclosure is a reaction to public pressure, it is likely that the company will discloses only the minimum required information (Cho and Roberts, 2010). However, if the information disclosed is seen as not sufficient, or if stakeholder demand increases over time, disclosure should improve over time (Cavana and Becker, 2016; Stanny, 2013). The power of the legitimacy argument is driven by the effective influence of the stakeholders/society on the company. Usually, media exposure represented by company size and company industry is used as a (weak) legitimacy proxy, while profitability is not connected to legitimacy arguments (Yunus et al., 2016).

Environmental disclosure is often explained by legitimacy arguments, as the carbon debate is relevant to the public debate (Patten, 1991, p. 305). The industry is relevant for the legitimacy discussion, as regulations, public debates, etc., are often connected with entire branches of industry (Wackernagel and Rees, 1997). Empirically, the legitimacy argument is supported by the analysis that carbon disclosure is heavily connected with the communication function, with limited engagement of the finance function. This is an indication that companies strive for external recognition but have no/limited interest in changing their behavior internally by managing actively their emission inventory (Windolph, 2013, pp. 37-49). The legitimacy argument is seen critically: as companies tend to disclose only positive information, disclosure is far from complete and very seldom audited (Andrew, 2013, p. 400, Blacconiere, 1994, Patten, 1992).

III.   Institutional theory

Institutional theory identifies three core factors additionally to the competitive factor that explain why actions/organizational structures of companies become homogeneous over time (CDP, 2015, pp. 2-3, DiMaggio, 1983, GRI, 2015, p. 2, Meyer, 1977). These factors are named isomorphic processes - coercive, mimetic, and normative (DiMaggio and Powell, 1983, p. 147).

Some authors use institutional theory (Campbell, 2007; Cooper and Owen, 2007; DiMaggio and Powell, 1983; Nikolaeva and Bicho, 2011) to explain the growth of sustainability and carbon reporting (Brown and Dillard, 2014; Cruz and Matsypura, 2009). Coercive isomorphism describes changes which companies go through when they react to external pressure. The argumentation of legitimacy theory already described the forces that pressure companies into disclosing carbon information. If these forces are putting pressure on some companies to disclose information, then, it is likely that companies with similar stakeholders will also disclose similar information (GRI, 2015, p. 2, Stubbs2014). Coercive isomorphism is fostered by the relevance/dependency of (single) stakeholder (groups), the homogeneity of the competition, and the homogeneity of the stakeholder groups, e.g. customers who pass requirements along to their suppliers (WBC, 2004, p. 25). This effect is officially fostered as standard setters explicitly recommend adopting commonly used practices (CDP et al., 2011). Additionally, uncertainty about the future actions of stakeholders, especially regulators, also support coercive isomorphism. However, uncertainty is more closely connected to mimetic isomorphism (DiMaggio and Powell, 1983).

Mimetic isomorphism refers to processes companies undertake in the face of uncertainty. The main argument is that companies (especially their upper-echelon decision body) emulate companies that they perceive as successful in phases of uncertainty (WBC, 2011a, p. 151). The blind-copy process of companies occurs regardless of whether this benefits the company or not. Uncertainty factors that might foster mimetic isomorphism could include uncertainty with regard to goals. While there seems to be consensus in science and society to limit global warming to two degrees Celsius, the implications for the business world broken down to each single company are widely uncertain. This uncertainty refers to the amount of required change in a certain industry sector, company and product line and the uncertainty when this change is likely to happen (CDP, 2015d; DIN Deutsches Institut für Normung, 2011; IPCC, 2014; UN2, 2015), even though sector ambitions do exist (CDP, 2015g). An example of this uncertainty based on different messages a company receives is the general target to stop using fossil fuels at some point of time in the future (Uni, 2015). In contrast to this governments still subsidize new investments in the coal and gas industry (Greshko et al., 2017; Robertson and Adani, 2016). This unaligned message brings uncertainty about the necessity to green investments. Another reason is uncertainty about the approach to reaching a particular target, e.g. if business models and technology are changing and different options require different decisions. Uncertainty about future actions by stakeholders & regulators - especially if stakeholders send signals that change will happen, but the road map is unclear (CDP, 2015d; ECOpoint Inc., 2015; UN2, 2015) - could also foster mimetic isomorphism, especially in a direction of respected best practices communicated by institutions such as the CDP or GRI.

Normative isomorphism processes also ensure that companies solve problems homogeneously over time. If people are taught similarly how to solve specific problems, it is likely that the problems will be approached similarly. This can occur as the result of formal education, specialization or socialization on the job. Employees characterised by high education levels, specialist job role profiles with typical career paths, or domination by education providers and consulting companies, thus tend to support normative isomorphism. These conformities are further supported by bureaucratic processes and regular interaction among specialists, e.g. at industry conferences or cross-company training activities (WBC, 2004, pp. 6-9, 147-154). Formally, this leads to more and more homogeneous actions by companies. Carbon disclosure by organizations should thus be expected to converge over time (Cormier et al., 2005; Luo et al., 2012; Matisoff et al., 2013; Matisoff, 2013).

However, the disclosed information does not necessarily represent reality, as these isomorphic processes also lead to a disconnection between the disclosing headquarters and the production site. Decentralised entities also develop strategies to avoid audits (DiMaggio and Powell, 1983, p. 25). Where carbon accounting is concerned, this implies that the risk of window dressing and green-washing increases over time (WBC, 2004).

IV.   Agency theory

The theoretical foundation that carbon disclosure is done as signaling argument in the agency theory is based on the assumption that some or at least a few financial capital providers associate green actions with positive financial performance (Clark et al., 2015; Endrikat et al., 2014; Figge and Hahn, 2002; Guenther, 2013; Niehues, 2014; Schaltegger, 2011, 2013). Additionally, some investors invest only in companies that pursue similar social, ethical or environmental aims. As transactions are dependent on sufficient conformity of values, this can also be a cause for the requirement of the respective signal (Hufschlag, 2008, pp. 56-73). As a consequence, investors view carbon disclosure as a cherry signal and associate disclosure with the existence of a climate-change strategy (IPCC, 2014, pp. 152-154, Meyer, 1977, p. 353, Rankin, 2011, WRI, 2015e). Using this as the underlying assumption, voluntarily reported carbon figures can be explained as signaling information (Miller and Plott, 1985; Verrecchia, 1983) designed to enable the investor to differentiate between cherries and lemons (Akerlof, 1970; Comyns et al., 2013).

Here, the disclosure is done to reduce the information asymmetry between companies and investors or other relevant stakeholder groups (Connelly,

2011, Hahn, 2015, pp. 86-87, Healy, 2001, p. 406). Sending a cherry signal to the investor which enables to distinguish the reporting company from a green-washing lemon reduces the cost of external financing (Gerke, 2005, p. 257, Hacker, 2001, p. 666, Haeseler, 2006, p. 122, Velte, 2008, p. 20). As companies try to promote an environmentally friendly image, companies in the role of agents disclose more information as part of their signaling to influence their stakeholders in their role as principals (Brouhle, 2010, pp. 521-548, Clarkson, 2011, Hahn, 2015, pp. 86-87). The empirical evidence that CEO statements in CSR reports lacking critical self-assessment, independent of their performance is an indication that the current messaging contains many signals from green washing lemons (Barkemeyer et al., 2014). If a company is requested to disclose information voluntarily, but does not disclose the relevant information, stakeholders are likely to interpret the this disclosure option as a signal that the company is striving for a situation in which stakeholders are not able to distinguish between lemons and cherries. As cherry companies have an incentive to avoid such a situation, cherry companies have the incentive to disclose relevant information. Considering that the environmental efficiency maturity of companies is not a pure cherry lemon separation but rather a range of different maturity levels, all companies above the lowest level of stakeholder interpretation do have an incentive to disclose some information voluntary to distinguish themselves to the lowest/lower performing companies (Healy and Palepu, 2001, p. 422). However, companies struggle to find the right amount of disclosure, as the value of the disclosure can be ascertained only after the information release, and standards can provide only minimal requirements (Ascui, 2011, pp. 579-605, Bowen, 2011, p. 376, Dhaliwal, 2009, pp. 56-60, Doda, 2015, pp. 2, 69).

Here, information intermediaries (e.g. analysts' ratings) aid the investors by indicating which information is missing and how to interpret the disclosed information. Ideally, these intermediaries examine all regulations/standards and point out the relevant information that fulfils the requirements to distinguish between lemons and cherries (Healy and Palepu, 2001, p. 407). Ideally, information requirements are chosen in a way that enables stakeholders to differentiate between environmentally friendly and non-friendly companies. This implies that the information must be honest and trustworthy (Hahn et al., 2015, pp. 86-87). To indicate to the stakeholders that the information is credible, the disclosing company can voluntarily engage a third-party intermediary that uses its reputation to confirm the correctness of the information (e.g. an audit company). Alternatively, the data can be validated over time, e.g. against prior disclosures, so that stakeholders can see this information as plausible (Healy and Palepu, 2001, p. 425). The audit argument justifies voluntary audits of carbon disclosures (Erlei, 2007, pp. 49-53, Healy, 2001, p. 415, Leftwich, 1983, pp. 23-40).

Ideally, the company finds signals that lemons cannot produce or that have significantly higher costs associated with the signal. E.g., in agency theory, long warranties for products are considered to be a signal for high product quality, as competitors producing with lower quality would face higher repair/replacement costs if they tried to imitate the warranty signal (Calza et al., 2014). Applied to carbon accounting, it is likely that the production cost of the disclosure is identical for all companies. Therefore other measures, such as environmental rankings, efficiency KPIs, etc., must be used to differentiate between cherries and lemons (Connelly, 2011, Krakel, 2010, pp. 28-34, Spence, 1973). Besides the costs for the message itself, there are also costs associated with ensuring the credibility of the information. If e.g. the carbon-efficient company discloses an audited carbon figure based on a generally accepted standard, the non-efficient company would be unable to send a similar signal to the market, as auditors would potentially refuse to confirm this number. The adoption of environmental certifications functions similarly (Castka and Balzarova, 2008; Teng et al., 2014). Conflicting with full disclosure is also the question of whether disclosure would include proprietary information. If the information disclosed implies a competitive disadvantage, the company will not disclose this information, even if this means an increase in the company's financing cost (Dye, 1986, pp. 331-336, Hayek, 1945, p. 519, Healy, 2001, pp. 423-424, Verrecchia, 1983, Wagenhofer, 1990a, 1990b). Looking at the way costs are distributed among stakeholders, it is clear that all disclosure costs have to be covered by the existing shareholders, while all stakeholders can potentially benefit from the information. It is thus likely that an underproduction of disclosure exists (Beaver, 1981; Healy and Palepu, 2001; Leftwich, 1980; Watts and Zimmerman, 1986)

V.   Upper-echelon motivations

Looking at the arguments for companies to disclose information, it is important that the decision for or against disclosure is taken by the management of a company. This upper ecchelon thus has a game-theoretic element to decide whether or not and how much to disclose GHG emissions. Looking at the motivations of the CEO, he or she can have monetary and non-monetary incentives to disclose or refuse to disclose information (Fabrizi et al., 2014, p. 311). The financial motivations of the CEO to disclose information are connected to the incentivation scheme, e.g. stock options incentivize over the long term, while short-term financial targets incentivize in the short term (Fabrizi et al., 2014).

Alongside this financial incentive for managers to disclose GHG information, managers have also non-financial incentives to disclose GHG information. The arguments can be summarized under personal values of the CEO, especially fiduciary obligations, management talent signaling, corporate control contest, and litigation cost hypothesis. In general, different CEOs have different values, preferences and perceptions of the future (Meyer, 1977, p. 58, Richter, 2003, p. 3). Therefore, the CEO's values shape ESG disclosure decisions based on his or her previous experiences (Cotter and Najah, 2012; Hahn et al., 2015; Jira and Toffel, 2013; Kolk et al., 2008). The ability to decide for disclosure is thus linked to the CEO's values, and to his or her standing, which enables him or her to act according to these values. An old/experienced CEO with a successful track record may want to give something back to the community, while a new CEO still unknown to the capital market faces potentially higher pressure to achieve short-term successes (Deegan, 2002, Fabrizi, 2014, pp. 311-326, Gray, 1995, p. 335, Hahn, 2013, Milne, 2002, Suchman, 1995). The viewpoint of the requirement to give something back is often seen as fiduciary obligation. In this case, the manager sees a responsibility to engage with climate change and analyses, considering which measures to take to reduce the negative effect of the company based on measures beneficial for the company in the long run (Fabrizi, 2014, Okereke, 2007, p. 480). The causal relationship between CEO values and GHG disclosure is not clear, however. As the shareholders decide who should be the CEO, the values of the CEO and his or her environmental viewpoint could be already a consideration in hiring a CEO (Hahn, 2015, pp. 89-93, Stanny, 2013, p. 197).

Another argument for voluntary disclosure based on CEO motivation is the management talent signaling hypothesis (Healy and Palepu, 2001, p. 4423). According to this, the disclosure shows shareholders and stakeholders that the CEO is already coping with upcoming challenges and therefore indicates that he or she is especially talented (Healy, 2001, pp. 423-424, Trueman, 1986, pp. 53-71).

Alternatively, disclosure can be explained by the corporate control contest hypothesis (Healy and Palepu, 2001, p. 421). Under this proposal, shareholders hold the CEO responsible for stock performance. Additionally, it is assumed that unexpected negative information has a higher impact on stock prices than negative effects that have been communicated early and explained. A GHG inventory can thus help manage shareholder and stakeholder expectations and reduce the risk of unexpected shocks to the stock price. These shocks are associated with high CEO turnover (Healy and Palepu, 2001, pp. 421-422). This argumentation can also be connected a the CEO's standing within the company. A CEO who is also owner of the company has less incentive to support this kind of disclosure, and departing CEO's also have less of an incentive than CEOs just starting out in the role (Fabrizi et al., 2014, pp.3111-326).

The last non-financial argument to explain a CEO's motivation for GHG disclosure is the LITIGATION COST HYPOTHESIS (Healy and Palepu, 2001, p. 423). The disclosure based on litigation arguments is connected to negative information and the managers' fear of either being held responsible for this information or facing penalties for late disclosure. As it is unclear if GHG emissions may lead to penalties in future, CEOs may have an incentive to disclose negative effects early, if it is likely that penalties will occur. If CEOs sense that an issue will land in court anyway, it may be beneficial to disclose the relevant information over several periods to minimize stock effects on certain dates, as these effects are often used to determine penalties. The litigation argument is negatively associated with forecasts, as managers can fear penalties from incorrect forecasts (Healy, 2001, p. 423-424, Skinner, 1993, pp. 407-443). If shareholders fear similar litigation issues, this can be solved in a manner similar to other principal-agent settings (Oberhofer and Dieplinger, 2014).

VI.   Theoretical overview of disclosure strategies

Based on the different theoretical motivations for voluntary disclosure the way on how information is disclosed varies. A common motivation for voluntary disclosure is, that in all cases the company expects that their stakeholders have a positive view on green measures and anticipate the positive relation of the pays-to-be-green business case. Figure 3.4 shows here the arguments for the pays-to-be-green business case on the left side. These arguments are associated with a causal relation between corporate environmental performance (CEP) and a positive corporate financial performance (CFP). The legitimacy theory argues from a reactive to pressure behavior, while the institutional theory highlights the reaction on competitors/peers and the agency theory

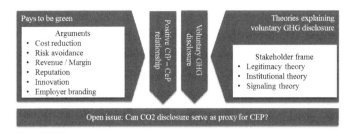

Figure 3.4: Explanation for voluntary GHG disclosure (Niehues et al., 2017).

focuses on proactive behavior of the reporting companies. In all cases the question arises, what information shall be disclosed which is perceived as green measure. One option here is the CO_2 inventory (Niehues et al., 2017).

The way how this CO_2 inventory is presented differs from the motivation why the company discloses this information.

Companies reacting to external pressure are likely to disclose only the amount of information which is required legally or defined as minimum expectation from powerful stakeholders with the aim to remain its license to operate as shown on the left side of Figure 3.5. This kind of disclosure can be theoretically explained by the legitimacy theory. Stakeholders with an information need which exceeds the disclosed information shall seek for options to increase the pressure on the company. A contrasting disclosure behavior is chosen by companies which want to send a green signal to its stakeholders with the aim to convince their stakeholders that the company is a cherry company in the field of green measures. Those companies voluntary disclose sufficient information to be seen as best in class company. This behavior can be explained by the agency theory and the pays-to-be-green argumentation. Stakeholders interested in additional information shall here communicate why the additional information is required and how the additional information can be used to distinguish between environmental friendly and less friendly companies. Companies following the institutional theory logic to copy companies they perceive as benchmark disclose similar information. As the best in class reporting companies are not necessarily those companies which are perceived as benchmark from the reporting companies, it is important for stakeholders to influence the perception on the benchmark and highlight the relevant elements of the disclosure behavior of the benchmark (Niehues and Dutzi, 2018).

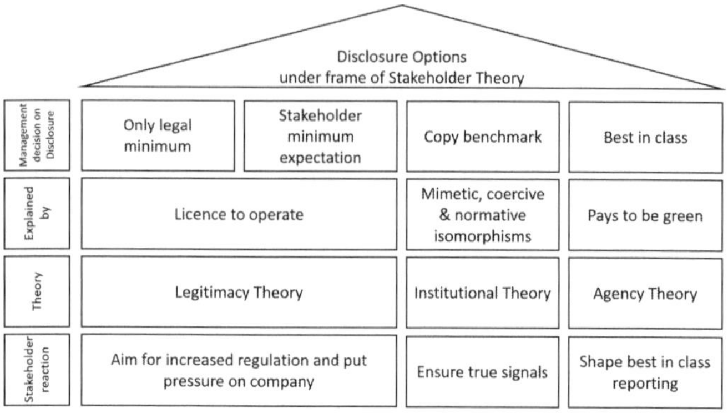

	Disclosure Options under frame of Stakeholder Theory			
Management decision on Disclosure	Only legal minimum	Stakeholder minimum expectation	Copy benchmark	Best in class
Explained by	Licence to operate		Mimetic, coercive & normative isomorphisms	Pays to be green
Theory	Legitimacy Theory		Institutional Theory	Agency Theory
Stakeholder reaction	Aim for increased regulation and put pressure on company		Ensure true signals	Shape best in class reporting

Figure 3.5: CO_2 disclosure options under the stakeholder umbrella (Niehues and Dutzi, 2018).

This review has shown that legitimacy theory, institutional theory, and agency theory can be applied to explain the existence of different voluntary CO_2 reporting strategies. Here, only agency and to some extend institutional theory intend to convince stakeholders to perceive the company in a manner that is as environmentally friendly as possible. To provide a CO_2 inventory, a sound understanding of various disciplines is required. This results from complexity in reporting along the entire value chain and the dynamic nature of the creation of CO_2 emissions. External analysts thus need detailed infor-mation to make judgements about the environmental impact of the reporting company and respective trends (Niehues and Dutzi, 2018).

Figure 3.5 summarizes the connection between the different disclosure strategies of companies and their theoretical foundation under the stakeholder theory umbrella. Even though this model is presented in a static way, the interactions between stakeholders, peer companies and the reporting company over time has an impact on the disclosure patterns over time (Niehues and Dutzi, 2018). From this theoretical overview it is relevant for stakeholders and companies how to identify high quality reporting.

E. Disclosure characteristics as indicator for disclosure strategies

I. Complexity in counting CO_2 emissions

GHG disclosure is generally done according to a certain reporting standard, e.g. the GHG Protocol (WBC, 2004, 2010), and reported either via CSR reports or other reporting institutions such as the CDP (Armstrong, 2011; CDP, 2015b,e). Reporting of carbon figures requires a sound understanding of the various tools and methods to measure carbon, which is the result of dynamic chemical and physical process. This implies different skills and hence different research areas. The different background of the experts in the fields of GHG accounting imply that the research agenda on carbon accounting has different priorities (Hambrick and Mason, 1984). The chemical process implies a scientific error in addition to the measurement error (IAA, 2011).

First, it is important to recap that measurement of CO_2 emissions is a dynamic process in which emissions need to be added over the full value chain, as the cumulative impact of carbon is relevant. Hence, the boundaries of a carbon report cover not only the boundaries of the activities of the reporting company (scope 1) and the energy used for production (scope 2) but also activities by other parties. This implies the supplier (including transportation), so-called SCOPE 3 UPSTREAM EMISSIONS (WBC, 2011c, p. 25), and, in a life cycle assessment of the product, the emissions caused during usage by the customer and continuing through disposal at the end of the life cycle, so-called SCOPE 3 DOWNSTREAM EMISSIONS (WBC, 2011c, p. 25). These scope 3 emissions often imply activities that cannot be measured by the reporting company. These emission figures need to be estimated, which results in an estimation error (Burritt et al., 2011; IAA, 2011).

This estimation error increases in significance based on several effects. One effect is the domination of scope 3 emissions over own emissions (Downie and Stubbs, 2011, 2012), another effect stems from constraints in the resources available to measure carbon emissions and the usage of extrapolation based on sampling (Goldsmith and Basak, 2001). In terms of calculation schemes, there are different options and systems available to calculate data. Depending on data availability, companies can choose different options that lead to significantly different results (Lewis et al., 2014; Milne and Grubnic, 2011; TNO et al., 2012; Auvinen et al., 2011b). The methods used to calculate CO_2 emissions are not defined everywhere (Andrew and Cortese, 2011a). However, there seems to be a strong correlation between the level of disclosure and carbon performance, which is a indicator against green-washing strategies (Luo and Tang, 2014). Looking back at the determinants of decision usefulness of information based on the concept of Dohrn (Dohrn, 2004), these limitations in data availability stand in conflict with other accounting principles, such as reliability, accuracy and comparability. Depending on the level of disclosure and the specification of the standard used, the principles DECOMPOSITION OF INFORMATION, CLARITY, and PREDICTION ELIGIBILITY are under risk as well. To date, only a limited number of research papers have addressed the causes of data-quality issues in carbon reporting. These have highlighted the complexity and uncertainty of GHG measures, technical issues, lack of transparency for stakeholders, and the limited reliability of disclosed information and its audits (Gray, 2010; Milne and Grubnic, 2011; Talbot and Boiral, 2013).

Looking at financial accounting standards and applying them to GHG accounting standards, it has to be taken into account that flows of financial activities are completely different from accounting for chemical processes. A financial Profit and Loss, a balance sheet and cash flow can be crosschecked against each other, and certain financial flows will always be tracked at the accounts. Comparing this with the landscape of GHG emissions, an observer would see a smoking chimney as output. From this starting point, GHG accountants can investigate how best to measure output emissions. Technically speaking, the most valid approach would probably be to measure all emissions escaping across all production steps in which fossil energy is burned, and to transfer this information into the accounting books. The discussion around proper measurement and defeat software in cars during the Volkswagen emissions scandal (Schnell, 2015] also turned public attention to measuring costs, different measurement methods and remaining errors. Currently it is unlikely that emissions quantities can be permanently measured for each chimney (which can also be a regular car). Therefore, more cost-efficient processes must be developed to measure emissions over the life cycle. A very simple model can be a linear model of input and output of relevant elements (CMU, 2013). At the very start of the life cycle, in which base materials are processed to another standard product, e.g. burning of coal to generate electric power, these calculations and conversions can be very accurate. As soon as the output and input factors are less granular and are measured in abstract elements such as cost, such models become less accurate. E.g. emissions per €1000 service fee in the service sector cannot be reflected in an accurate CO_2 number. Even more precise methods to estimate emissions calculate different numbers. In the transport area, for instance, different systems calculate different CO_2 emissions based on the same input data, as underlying assumptions and methods are not harmonized. Depending on these decisions, carbon emissions will result in different figures. Unlike in finance, these differences are relatively rarely discussed in the research to date (Lewis et al., 2014; Milne and Grubnic, 2011; TNO et al., 2012; Auvinen et al., 2011b). As soon as indirect measures are retrieved, other errors follow, as conversion factors are often calculated for a specific geographical region and a certain technical standard. With lots of companies operating in different geographical regions and changing their production facilities dynamically, the conversion factors used should anticipate this as well. Moreover, hidden hazards of emissions are also not always reflected in emission factors (Goldsmith, 2001, Hartmann, 2013, p. 548). All of these rather technical issues grow more complex on the measurement/accounting side, as these emissions have to be collected over the entire value chain, which includes many legally distinct entities. Therefore, the next section describes the measurement error involved.

II.   Measurement error

The measurement error of a carbon inventory is the product of a list of different factors. Most important is the limited availability of the emission sources. The majority of emissions in a production system are usually caused by an entity other than the reporting entity. The reporting entity has only a limited ability to gain full information about these indirect carbon emissions. The next big issue concerns differences in understanding of scoping. Most industries currently lack a consensus as to which emissions have to be accounted for and which left out. For a single company disclosure without indirect emissions, this deficit can be easily compensated for by disclosing the respective underlying definitions. In a multi-step value chain, on the other hand, it is extremely complex to ensure that all involved parties follow the same scoping (Andrew and Cortese, 2011b; Downie and Stubbs, 2012; Hahn et al., 2015). Leaving out these two main issues for consistent reporting, there are still a number of operational challenges that must be met in order to measure a correct carbon number. Starting with the limitation that not all carbon-emitting processes can be measured and tracked permanently with limited resources, some sampling procedure is required. This leads to measurement errors in sampling. The more complex a company's portfolio and value chains (resulting in a higher standard error), and the fewer sampling points, the higher the margin of statistical error. As the expectations of regulators and stakeholders vary, there might be a significant number of companies opting for rather cheap but less effective sampling strategies (Goldsmith, 2001, Milne, 2011, Hartmann, 2013, p. 548). These different expectations might also impact staff experience with carbon numbers. Ideally, the staff that is responsible for carbon reporting should be qualified to judge the accounting process, social impact and ecological impact. The example of accounting for an afforestation project shows that different standards lead to completely different judgments of carbon offset for the same project (Lee et al., 2013, pp. 53-62). As chemical, biological, ethical and accounting research streams have different priorities for accounting, the accounting principles and their priorities are weighted differently by organizational groups. Companies' environmental accountants may have to choose to prioritize between different accounting principles facing varying recommendations (Bowen and Wittneben, 2011). Another issue is that carbon accounting measures output only. A check for completeness via the connection between P&L and balance sheet is not possible. The same applies for missing completeness checks via cash flow statements (Andrew and Cortese, 2011b; Green and Li, 2012; IAA, 2011; Talbot and Boiral, 2013). This lack of accounting logic has to be added to the accounting challenges known from financial accounting, e.g. the question of data collection systems, transferring units, allocating the emissions to the correct reporting period, etc. It is argued that this might not be challenging, as there are standard processes available in finance; however, either operational data collection processes have to be used to collect these data at a finance level of quality, or else finance systems/processes have to be enabled to collect operational data (Martinov-Bennie, 2012). These points all act in combination and increase the margin of error. To reduce the error margin and thus increase the robustness of the carbon inventory and its credibility, some companies voluntarily seek an audit. The next section describes the limitations of the audited GHG inventory.

III.   Audit expectation gap

To increase the reliability of the disclosure and ensure data quality for management purposes, companies can seek external verification (IAA, 2011). Understanding the complexity of a carbon report helps to understand the limitations an auditor faces when tasked to assure the correctness of the data. Unfortunately, audits are often completely missing for voluntarily reported information, this can be interpreted critically for the trustworthiness of the report (Mock et al., 2013). Yet even after carbon figures have been externally verified, there are still concerns about data quality and comparability of carbon figures and KPIs. These concerns negatively affect the usefulness of the reports to decision-makers. This is especially important as regulators e.g. in Australia and the European Union depend on carbon disclosure for their regulation approaches (McNicholas and Windsor, 2011). This chapter gives an overview of the complexity of carbon numbers and the limitations in terms of accuracy and comparability. If an audit has been performed, the question still remains of the audit scope. It must be ensured that the audit scope covers all disclosed emissions and not just portion, e.g. only scope 1 or 2 (Niehues and Dutzi, 2016). The next element that should be checked to interpret the trustworthiness of the audit is the level of assurance. Companies can opt for limited assurance instead of reasonable assurance, as it is mandatory for financial reporting. The usual assurance statement for CSR reports is limited assurance. The objective of a limited assurance statement is only a negative statement that the auditor has not found any indication *that the GHG statement is not prepared, in all material respects, in accordance with the applicable criteria* (IAA, 2011, p. 7).

Looking at the procedures required of the auditor aiming for limited assurance, the auditor is only required to review the control environment, the business processes, the entity's risk assessment and the information system. Not required is to evaluate and monitor the control system, test it and evaluate the sampling and estimation methods (IAA, 2011). These differences show that limited assurance has significant limitations, and that the trust gained for this audit statement should also reflect these limitations. Coming to the gained trust, some studies state that the trustworthiness of a CSR/carbon report increases if it is assured by a (reliable) auditor (Wong and Millington, 2014). However, readers should take into account that auditors see a growing market in environmental assurance services, and as the audit is often voluntary, the potential conflict on the part of the auditor should be acknowledged (Borkowski et al., 2011). Even though auditors implement processes to limit the effect of these conflicting interest' on assurance quality (O'Dwyer et al., 2011), these limitations of required audit performance and the competition between audit companies often lead to an audit expectation gap (Best et al., 2001; Green and Li, 2012; Green and Zhou, 2013; Olson, 2010). Or, if the audit service does not add enough trust, release of CSR reports will be ignored by the capital market (Cho et al., 2014). Generally very few research studies have been released in the area of CSR assurance with its determinants (Berthelot and Robert, 2011; Rankin et al., 2011), outcomes and assurance practice (Green and Taylor, 2013; Green and Zhou, 2013). As a result, lots of research gaps remain in the area of voluntary GHG emission reporting (Hahn et al., 2015, p. 93). After seeing how complex it is to calculate and verify a carbon inventory, the next section describes how and whether external stakeholders can use the carbon information to compare companies - along with the limitations that should be taken into account.

IV.   Ensuring a complete carbon footprint

Understanding the complexity of measuring CO_2 emissions is a precondition to understanding whether two companies can be compared based on their carbon inventory. The challenges involved in setting up just one carbon footprint for a single company suggest that a carbon number should probably not be used as the one and only single indicator but rather as one indicator among several.

The challenges and prerequisites that should be taken into account in the effort to compare companies based on their carbon inventory could be seen as the requirement to compare APPLES WITH APPLES. A like-for-like comparison starts with a description of using similar reporting elements, scopes and standards. Afterwards, challenges involved in finding similar companies are reviewed, before the potential KPIs in theory and practice are shown.

As almost all reporting standards highlight the importance of setting reasonable boundaries and disclosing the respective scope, it seems obvious that a comparison has to be done on identical/similar boundaries and that a report should stick with consistent boundaries applied to the whole company. Unexpectedly, a study analyzing the FTSE 350 reports in the years 2010-2013 found out that nineteen percent of the companies failed to describe how they defined their organizational boundaries, even though this a legal requirement (Kasim et al., 2016, p. 11). Consistent with these results is a study covering the years 2005-2009, showing that only fifteen percent of the companies reported a complete GHG inventory (Liesen et al., 2014). Missing data is most often the result of omitting some locations, defining some process steps as beyond a company's operational control, leaving out refrigerants, or setting a materiality level under which emissions are not reported. Additionally, sometimes companies use different reporting periods for financial accounting versus environmental reporting, which makes plausibility checks over time more difficult (Kasim et al., 2016, pp. 11, 13, 30). In terms of trending, the oil industry did not improve their GHG reporting between 1998-2010 (Chi et al.,

2013). This empirical evidence shows the importance of also actually quality-checking rather SIMPLE requirements of GHG reporting to ensure that carbon reports are comparable before analyzing the data quantitatively. Further issues can be explained based on the three different scopes, own emissions (scope 1), the emissions generated from energy used (scope 2) and emissions from subcontracted goods & services (scope 3 upstream) or emissions caused during the life cycle of product usage (scope 3 downstream) (WBC, 2004, p. 25). If companies want to reduce scope 1 emissions, probably the quickest and easiest solution is to identify the most carbon-intensive process steps and outsource them to a third-party subcontractor. This outsourcing step has no environmental effect and does not change a complete inventory, but it massively reduces the scope 1 emissions. So if targets/reports include only scope 1 emissions, it is important to take the vertical integration of production and production steps into account as well. Otherwise it is hard to analyze the environmental impact. While this idea sounds simple, some research papers argue that across industries only scope 1 and potentially scope 2 emissions can be compared due to data quality (Andrew and Cortese, 2011b; Hahn et al., 2015). The next element are the scope 2 emissions. These emissions are driven by two factors. One factor depends on total energy usage, which is mainly electricity. The second factor depends on the energy mix either on a country level or on a provider level. So a low scope 2 carbon number can be the result of different effects. One possibility is that the company operates very energy-efficiently and thus has only a low scope 2 carbon footprint. The second option is that the company uses a green tariff and pays a premium for low carbon emissions. Or perhaps the company operates in a country with carbon-efficient energy suppliers, among which nuclear power is also ranked from an emissions standpoint. These three different causes should be judged separately by questions around the ability to produce very efficient. A total scope 2 carbon figure cannot be used to distinguish the causes. Here the supplementary KPI energy usage and green energy usage could close some information gaps.

The real complications start with scope 3 reporting. The core issue is the fact that carbon accounting has a significantly greater scope than financial reporting. Accounting of upstream and downstream emissions can be compared with the complexity of accounting for warranties and off-balance sheet financing options (WBC, 2004, 2011a). Scope 3 emissions are all emissions FROM CRADLE TO GATE (Schmidt, 2009) are not covered under scopes 1 & 2 emissions. These can be further classified into categories.

To judge whether a carbon inventory is complete, the analyst must know which categories are relevant for which company. Typically, each sector has special categories that dominate the inventory. Even though scope 3 accounts for most of the emissions and is highly important to an understanding of the full picture, a majority of companies of the FTSE 350 do not disclose their scope 3 emissions. Therefore, the CDSB recommends that regulators request more mandatory disclosure. This disclosure should be defined clearly by standards to ensure comparability. This would benefit all stakeholders, as definitions of mandatory emissions are more transparent (Kasim et al., 2016, pp. 6, 11, 30).

The next step to ensure like-for-like comparability is to determine whether the different collection methods bring comparable results. As big multinational corporations often have a number of suppliers in the five digits, and as even these subcontractors need to collect data from their value-creation process, companies have to carefully decide how to collect data. In many cases, subcontractors will also refuse to report carbon data, as these data often contain proprietary information. Alternatively, subcontractors could opt to provide favorable but incorrect data. Further reasons for incorrect/inconsistent data could result from the usage of different settings, as perhaps not all subcontractors have the expertise required to provide a correct carbon footprint. This can motivate companies to calculate their carbon footprint based on industry averages or other estimation methods. If two different companies use different sources/assumptions for industry standards, the results can differ significantly even though the underlying business is identical.

Even if all subcontractors provide reasonably good data, it has to be considered that, in addition to creating a carbon inventory, they also still face issues of performing a product/customer calculation. The example of a milk-run allocation shows how complex it is to allocate emissions generated in production to the respective products (Appel et al., 2010; WBC, 2011b). These allocations might be more complex in *Verbund production* (ChemCologne, 2016); this is especially common in industrial parks in the German chemical industry. Furthermore, scope 3 emissions are subject to external verification less often, which increases the risk of error (Busch, 2011, Kennedy, 2015, pp. 17-19).

These boundary issues could be solved by ensuring a common understanding between industries based on clear standards which reference specific industry standards. These industry standards need to define how organizational boundaries must be determined. Therefore, the next section describes the issues around using the right standard.

V. Reporting Standard

There are various stakeholders involved, publishing different reporting standards indicating how the report must be drawn up. For carbon reporting, the CDP lists 72 standards for emission calculation (CDP, 2016c, pp. 94-96), and for verification of GHG emissions around 40 acceptable standards (CDP, 2016b). These are only the standards identified by the CDP as being acceptable without judging or comparing them. Additionally, there is an endless number of standards which fail to meet the CDP minimum requirements for reporting/verification standards. Therefore, stakeholders who are interested in judging if the companies aims for comparability should be able to understand the reasons for different standards and the respective stakeholders, and to obtain guidance on how to identify the specifics for a respective company to enable a judgment as to whether or not a company uses a standard that meets the requirements. Globally, the GHG Protocol seems to be the dominant standard on calculating emissions. As two documents (WBC, 2004, 2011a) are covering all sectors, industries, legal and geographical locations, this standard is formulated in rather broad terms. To cover the named uncertainties of the carbon reporting, more specific industry standards must to be developed to cover industry-specific priorities and measures (Eccles and Serafeim, 2012). However, there is still the outstanding challenge that these specific and high industry standards are often not applied by the reporting companies in keeping with the general principles of the standard (Dragomir, 2012). Sometimes, these specific standards are even ignored completely (Eccles and Serafeim, 2012). Starting with this global overview of standards that become more specific, some institutions also release standards to fit their needs. This may be to ensure identical information over a certain regulatory area (Greenhouse Gas Division Environment Canada, 2004) or because an industry-driven NGO aims to invent a sector standard, as only a limited number of sectors actually have standards (Greene and Lewis, 2016). Other standard-setters e.g. the IFRS, the IDW, or the ISO institutions, also publish standards of their own, as carbon reporting often forms part of other reports, e.g. integrated reporting (Brown and Dillard, 2014; Cheng et al., 2014; The, 2013a; Leo et al., 2011). As there are reasons to release own standards of one's own from a standard-setter's perspective, it has to be ensured that the content of these reports is as comparable as possible, and that reporting companies cannot simply apply a poor standard to maximize their ability to choose and minimize comparability to other companies (Hahn et al., 2015, p. 93). With this aim, some global institutions try to re-harmonize the reporting. The

CDSB brings together the standard-setters (Kasim et al., 2016), while the CDP supports transparency over which companies use which standard. Next to ensuring similar reporting conditions it is important to identify relevant KPIs to rank companies.

VI.   KPIs

There are several options to identify leading companies in a special field. Besides rankings (e.g. the DJSI, CDP Score, etc.), quantitative assessments based on KPIs can be performed as well. If quantitative KPIs are chosen, the results have to be interpreted. When interpreting KPIs, it is important to understand the common KPIs and their trends within the industry sector. In financial accounting, big finance intermediaries classify all companies to respective industry and sub-industry sectors. E.g., Bloomberg uses BICS, while MSCI and Standard & Poor define GICS. For GHG emissions, because these emissions are not (publicly) available, typically financial industry sectors are used by investors and other stakeholders. Based on the example of oil transportation companies, this missing classification can lead to interesting results. E.g., Company A uses oil tankers to transport oil, while Company B uses pipelines; the financial impact on oil changes, political crises, etc., might have a similar impact. However, the carbon footprint and the potential development of this footprint over time could be completely different.

Sector reporting standards could solve part of this problem, but such standards are currently at a rather early stage of development (Allison et al., 2012; Downie and Stubbs, 2012; Eccles and Serafeim, 2012). This is especially interesting as institutional investors are driving carbon disclosure (Cotter and Najah, 2012). Besides the industry sector, the interpretation of KPIs also has to be performed on the basis of a certain target. One aim could be to understand which company is on a road to two-degree compliance. In this case, the question has to be reviewed based on the production of products but also downstream along the product portfolio. Looking into product portfolios, the analysis extends to life cycle assessments (LCA). In some cases, e.g. the car industry, these downstream emissions can be compared based on accepted industry standards; in other areas, e.g. a bottle of shampoo, the carbon footprint is dependent on the length and temperature of showering activity, which is potentially not a relevant KPI for a shampoo producer (Carnegie Mellon University, 2016; CMU, 2013; Green Design Institute, 2007).

This is especially interesting as the two-degree target is a forward-looking approach and has ambitious industry targets including absolute emissions and expected industry growth. The general aim of scientific targets for companies is to use the global-warming scenarios of the IPCC for 1.5- or 2-degree warming and to use this scenario as a basis for a total global carbon budget. This global budget is then phased to the respective years and broken down by sector/region. Based on the sector assumption for general growth within the sector/regions for each sector output, stated in financial terms, a certain carbon budget and carbon efficiency target is calculated by the science based target initiative (based targets, 2016). Companies that set their own targets can use different calculation options and pathways to calculate a carbon budget designed to fit into the overall two-degree framework. The institution of science-based targets, which collaborates with the institutions CDP, WWF, UN global compact and WRI, offers seven different options (based targets, 2016).

While environmental stakeholders as the science based investment community or NGOs like the WWF broadcast stories of companies that set science-based targets (Forster, 2016; WRI, 2016), as of August 2016 the prevalence of these science-based targets is not very high, with 176 companies (Sci, 2016). Therefore, researchers criticize that the general targets of companies are not aligned with the IPCC target (Nicholls et al., 2015, pp. 4-10), and that in relevant sectors, e.g. the car industry, no company has set a science-based target (Baker, 2016). Besides companies setting themselves science-based targets, it is also important that the required change finds a sufficient amount of capital. This can be done by investors who focus on investment criteria aligned with the scientific targets (Wor, 2013). Investors have different options to support the aim of the two-degree target. A direct way is to invest only in companies in which the value-creation process aims to create a carbon-neutral society. Examples are clean-energy investments, e.g. a solar park, or the producer of such technology, e.g. a windmill producer. However, if the portfolio is invested solely based on these very clear criteria, there are two main issues. From the risk perspective, the investor is limited to a handful of sectors, while from the environmental point of view, only a limited number of sectors receive climate-relevant financing (Hoehne et al., 2015b,a; Thijssens et al., 2015; Weischer et al., 2016). Other options are to set up guidelines on how to ensure that low-carbon investments get credits and/or existing companies get motivated to invest their resources in low-carbon investments or focus directly on companies with a relatively low carbon footprint (Hoehne et al., 2015b, p. 34). Alternatively, investors can choose only companies for which carbon development is expected to meet the two-degree target. In the absence of company targets, forecast must be performed based on past performance, which, in turn, is dependent on voluntary disclosure (2de, 2016). Due to the lack of clear regulatory guidance and the various different approaches to investing, many investors are hesitant to act more decisively (Cochran et al., 2015). Therefore, sustainable investors are appealing to regulators to increase regulations on transparency for environmental issues as a way to supply the investment community with more reliable evidence (Thomae et al., 2016, p. 3).

In the absence of clear two-degree investment options, and given limited data availability/reliability of the environmental impacts of the portfolio, stakeholders can use existing environmental disclosures to identify hot spots within a sector, a portfolio or a technology. In these cases, rather rough CO_2 numbers help to set priorities for further investigation.

The KPIs used in research are often blamed as poor proxies to measure sustainability. Potentially due to the ease of availability, researchers often use either a sustainability index as a proxy (e.g. the DJSI or KLD ratings), or CDP scores (Krause and June, 2015). Alternatively, researchers often use the availability of information to measure sustainability. This implies variables such as whether a report, e.g. GRI report, was published or not, without examining the outcome of the quantitative measures undertaken (Dilling, 2010), while ignoring the quality of the information given, which varies greatly (Guidry, Ronald P., Patten, 2010; Patten, 2015). Worse are other measures, such as number of pages in CSR reports, etc. which cannot be associated with any environmental performance (Dhaliwal et al., 2012, pp. 723-760). Reliance upon these KPIs causes several issues. To begin with, membership in a certain index (or certain environmental score) is often connected with other characteristics of the company; research should account for these moderating effects. Another point is that indices are also not immune to criticism, as indices often rank companies simply based on the fact that these companies voluntarily disclose lots of information (Healy and Palepu, 2001, p. 427).

As numerous publications point out, however, the amount of information disclosed is also a function of stakeholder pressure to disclose environmental facts (Cavana and Becker, 2016; Cho and Patten, 2007; Deegan et al., 2002; Hahn and Luelfs, 2013; Milne and Patten, 2002a; Patten, 1991). Additional ratings face the same quality issues for quantitative measures (Windolph, 2013, p. 81). Given this combination, companies with non-sustainable records are nonetheless often included in sustainable leadership indices (Cho et al., 2012).

In light of this, some stakeholders, investors in particular, have developed quantitative measures of their own to define green companies. Low carbon indices, for example, are advertised to enable investors to reduce their carbon risk portfolio (MSC, 2015a, 2016; limited and Limited, 2016). The general approach here is that the investor invests only in companies with a carbon KPI that is lower than the mean/median (or other statistical threshold) within the respective industry. To judge whether the KPI is a reliable measure, the carbon data used must be assumed to be accurate and complete. Another underlying assumption is that it is reasonable to compare the product portfolio within a particular sector. The KPI in those definitions is usually CO_2 divided by either revenue or number of employees. In rare cases, a sector-specific output (e.g. number of cars produced) is also available (and even less often used). An explaining factor could be that the disclosure is often done on a very high level, and that interpretability is lost during data aggregation (Hartmann et al., 2013, p. 548). CO_2 is usually defined as either scope 1 emissions or scopes 1 & 2 emissions. This implies that outsourcing activities, or business models with a relatively short value chain, are automatically associated with a good carbon KPI. Alternative definitions for value add for a company are seldom used (San-salvador-del valle and Gómez-bezares, 2016, p. 29). The concept of using carbon intensity figures is also heavily criticized, as it does not reflect carbon risk (limited correlation) and does not reflect scope 3 (Lucas-Leclin and 2 degree Investing Initiative, 2015, pp. 2-16).

As the measures used to identify green companies by stakeholders and researchers are often poor, the question arises why only poor measure exists and if this is caused by missing regulations, poor reporting or a complex setting. The FTSE 30 companies are legally required to integrate environmental KPIs into their annual reports. Looking into an assessment of the publications, 90% of the companies identified CO_2 efficiency as a core risk of their company; however, only 27% (only 23% incl. assurance) of the companies use carbon efficiency as a strategic steering KPI (Kasim et al., 2016, pp. 5, 18-19). From these 27% of companies which use environmental KPIs, almost all used GHG emissions with a normalizing factor, mostly revenue or number of employees. Only 15% used more than one efficiency metric KPI (Kasim et al., 2016, pp. 11-12). This could explain why investors also use revenue, as the reporting companies fail to provide a better context, or an industry-accepted benchmark (Kasim et al., 2016, p. 5). One indicator of a lack of comparability over time is that 45% of the companies did not disclose at least one previous year (Kasim et al., 2016, p. 32). These rather poor results on data availability and lack of comparability can run the risk that even poor performers will claim to be high-performing companies and sent as respective signals to the market to gain free-rider rents.

F.   Model to explain disclosure strategies by disclosure characteristics

I.   Process flow between principals and agents

Combining the overview of different theories explaining voluntary disclosure into an agency setting, Figure 3.6 shows that company stakeholders from the general public have in addition to their options to approach the company directly the ability to influence regulators and investors to impact the company for more voluntary disclosure. This ability supports the legitimacy arguments within the theoretical construct. This pressure has to ensure that the disclosed information and the respective green measures are actually supporting green measures. Here finance intermediaries such as green investment experts and information intermediaries, e.g auditors can support the intentions of the stakeholders on how to influence the public.

  To explain the game element that managers and investors face, an agency theory information game between the companies and the stakeholders can be set up. To simplify the model, it can be assumed that the actors consist of a non-investor stakeholders, investors, and reporting companies, all homogeneous within their group. The number of potential signals are clustered to four different options plus the option 0, meaning no signal is sent to the market. The other categories are:

1.   Meet the minimum requirement that stakeholders or investors request. E.g. the investor signs the CDP, the CDP questionnaire is answered SOMEHOW. If regulators ask for transparency, the minimum requirement is met. This option also exists for investors, e.g. if stakeholders request investors to include environmental information in their decision-making process, investors could sign the CDP, but not look into data, as the published signature is sufficient to their requirements. The same principles could also apply to regulators, customers, employees or other stakeholders that can answer say, yes, we want the green option, if asked, but fail to include green elements in the decision-making process.

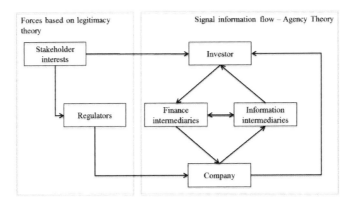

Figure 3.6: Environmental disclosure in an agency setting with the investor as principal and company as agent

2. Option two is the TRUE AND FAIR VIEW, which is a challenging approach incorporating the complexity of the reporting requirements (WBC, 2004, p. 6) (Zvezdov, 2011). This includes honest and fair reporting about the status of the green agenda. Signal two implies that the company does not have an outstanding green agenda and is not able to differentiate based on their disclosure. Most standards look for this fair view (e.g. IIRC, GHGP, CDSB, GRI4); however, there are indications that voluntary reporting fails to include negative information, as managers can choose what to disclose (Prado-Lorenzo and Garcia-Sanchez, 2010; Schaltegger et al., 2015).

3. The third option is for the actor to behave green and send the respective signal to the market. As in option two, this signal includes a true and fair view, however in option three the underlaying performance is the performance of a good performer.

4. The fourth option is the green-washing option, in which the signal of company in option three is copied at minimum cost, with a focus on the signal, not on the implication for the business. Examples might include reporting only non-critical information not directly comparable to peer companies (Hrasky, 2012), reporting different numbers over different channels, and reducing the numbers by reducing the scope or defining standards of the company's own (Depoers et al., 2014; Font et al., 2012).

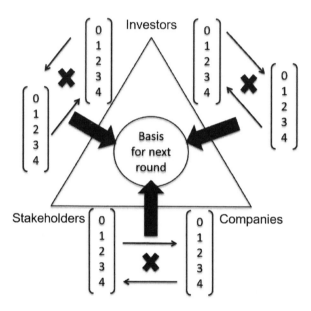

Figure 3.7: Signal flow and interpretation between investors, companies and stakeholders.

Figure 3.7 shows the different options and indicates that each agent sends signals and receives signals from its principal. As the signaling and interpretation process is happening all the time, this game can be set up with infinite repetitions. Depending on the respective signal from the other actors, each actor can react accordingly. In the given sample, the matrix contains 25 options at any given time for each stakeholder communication setting. This setting has to be multiplied by the number of stakeholders, as e.g. investors can communicate to the general public that they are focused on environmental investments but communicate their investment priorities differently to the companies in which they invest. With the described three stakeholder groups, three different matrices which have each 25 options can occur. As this situation is not necessarily a stable setting over time, the situation can change in future periods. After one iteration each of the three actors has the option to choose one out of five ways to react to the given status. Assuming that the three groups do not act homogeneously, as companies have different levels of environmental maturity, the public has a range of opinions and investor communities have different perceptions of environmental topics. Assuming that the iterations and the speed of reactions differ between and within these described groups, endless combinations of settings and developments are possible.

For simplification, only a few sample constellations and the respective implications for an iteration are described. An example between the investor and the reporting company could be that the investor sends signal 1 to the market that all investments should make their CO_2 emissions transparent and therefore sign the CDP. The reporting company receives this signal and has now the 5 different options. If the reporting company chooses 0, there is a significant risk that the investor will use his or her sanctioning power. Therefore, the company will most likely report at least the minimum required information (option 1). The company could also choose any other option than option 1; however it is likely that these options will involve a higher associated cost.

Whether the data reported is useful or intended to be used will be seen in the next iteration, which gives the investor the option of increasing his or her signal, e.g. by requesting more detail on the number or by not sending any additional information, which can be interpreted that the response by the company was sufficient. As Figure 3.7 of the agency information flow shows, the investor is not only principal and the company agent, but the investor is also dependent on society. Therefore, in terms of green signaling, the investor is also an agent while the society in which he or she operates is the principal. In this case, stakeholder groups could send signals to the investor that the investor community will lose their license to operate if the investors do not join the fight against climate change. As joining the fight against climate change might be perceived as rather abstract, investors could search for a transparent and cost-efficient way to demonstrate that they are doing something. Joining an initiative like the CDP by signing a letter of intent could be an option. This would explain option 1.

The next iteration would show whether option one is efficient to the stakeholders or not. If it is sufficient, an investor who does not agree with the pays-to-be-green argument might choose option 1 (Michelon et al., 2015).

One condition for green signaling is that the 'good performing' companies can send signals that the 'bad performing' companies cannot copy, or can copy only under relatively higher costs, and that these signals enable stakeholders to differentiate between cherries and lemons. The limitations in comparability and current usage of the respective KPIs indicate that the signals cannot be easily used to differentiate between good and bad performing companies. As the ability to report carbon numbers is likely to be equally efficient for all companies (or at least independent of a company's efficiency), the content of the report forms the criteria that distinguish between efficient and non-efficient companies. Therefore it has to be reviewed what disclosure strategy a non efficient company would choose to gain the stakeholders perception of being a green company (Erlei, 2007, pp. 148-198, Richter, 2003).

II.   Disclosure as manipulative signal

There is significantly clear evidence that companies try to use environmental disclosure as a tool to influence stakeholders. This begins with the fact that the wording is used in a percussive way in environmental reports (Arena et al., 2014). The information itself is unbalanced and focuses only on success elements instead of focusing on materiality. At the same time, the quality of information for material issues is poor (Comyns and Figge, 2015; Comyns et al., 2013). In other words, companies fail to report sustainability issues in their sustainability reports (Gray, 2006). Great examples for clear free-rider activities are studies that green disclosure is disconnected to emission performance (Chi, 2013, Doda, 2015, Kennedy, 2015, pp. 17-19, Krause, 2015). Even worse, disclosure sometimes is negatively associated with carbon performance (Cho and Patten, 2007; Patten, 2002). Using this negative relationship as an argument, disclosure activity and membership in an environmental leadership group rated mainly by disclosure could lead to fewer environmental actions (Freedman, 2004, p. 40, Patten, 2015, p. 46). This evidence can explain that capital markets currently do not see environmental disclosure as sufficient for decision-making (Andrew, 2013, p. 403, Ascui, 2014, pp. 7-10, Haigh, 2011, Solomon, 2011). Therefore, it is interesting to understand how markets react to environmental disclosure today. Generally, it is hard to measure how investors perceive voluntary environmental disclosures. The total limit of interactions observed by the company, however, indicates a rather symbolic reaction by stakeholders (Rodrigue, 2014). In a game study with business students, qualitative disclosure had a greater impact on decisions than quantitative measures (Rikhardsson and Holm, 2008).

Where financial market reactions are concerned, on average the release of sustainability reports does not have a relevant impact (Lackmann, 2010). This finding is consistent with the findings suggesting that lots of free-rider signals are sent within the sustainability reports. In terms of detailed market reactions, the companies with high reporting quality also derive significantly better market reactions. This result might suggest that capital markets are somehow able to assess the quality of sustainability reporting (Guidry, Ronald P., Patten, 2010). To increase the reliability and relevance of ESG reporting, some stakeholders would like to link ESG reporting and financial reporting in an integrated report.

The aim to integrate environmental information into annual reports within the integrated reporting framework is discussed controversial. This starts with the fact that huge lobby groups for integrated reporting do not come from environmental groups but from multinational companies and audit companies, which also have a huge influence on standard-setters and regulators (Andrew and Cortese, 2013). From an environmental standpoint, the missing link between climate strategies in integrated reports and the scientific two-degree target is criticized (Doda et al., 2015). From an investor's perspective, the missing link between environmental actions and company strategy and financial risk assessment is seen negatively, even though the integrated reporting standard aims to close the gap (Potter et al., 2013; Thomson, 2015). From an agency theory point of view, this missing link between environmental information and financial implications is expected, as the link discloses proprietary information which will not be voluntarily disclosed as a market signal. From an information perspective, it is not clear how the integrated reporting framework provides a guide to greater information comparability. To date, integrated reporting has not produced new reporting practices (Stubbs and Higgins, 2014) and the integrated reports for DAX companies do not include comparable information; the quality is very wide-ranging, and self-critical observations are missing (Bergius, 2015). As a result, voluntary application of the IR framework seems unable to close the information gap. However, it could help if the reports were standardized, if integrated reporting would have to follow GRI criteria, so that at least the structure would be comparable and CO_2 disclosure would have to follow defined sector guidance, which could help to standardize disclosure frameworks. If IR also have to fulfill the audit requirements as financial reports, carbon reports could close information gaps for systems and audits to increase the robustness of data (Bennett et al., 2013; Zvezdov, 2011).

Research shows that disclosing information to gain an environmentally caring image without actually acting in an environmentally friendly manner is prevalent. The challenges involved in integrating scope 3 emissions into comparisons are used as an argument to focus on a company's own emissions; this disables the validity of any relative KPI and disables total trending without adjusting for outsourcing/insourcing activities. These limitations narrow investors' use of GHG disclosures in decision-making. Therefore, CSR reports are not favourably perceived by investors, and divestment decisions are taken based on sector exposure.

III.   Research gaps in the agency model

Considering these often occurring green washing strategies into account the ESG accounting literature has very limited practical implications. One aspect is that the accounting literature focuses on the outcome and output of GHG disclosure but does not aid practitioners in showing how to interpret GHG inventory numbers (Hahn et al., 2015, pp. 93-95). Also missing is discussion of the question of which carbon-management activities are helpful for companies / what the success factors are for carbon disclosure (Kim and Lyon, 2011, p. 33). These findings from various literature reviews are especially interesting, as several researchers argue that the data quality is poor. Still, researchers prefer to analyze quantitative data settings rather than review the underlying data (Sullivan and Gouldson, 2012).

Transferring this finding into an agency model as shown in Figure 3.7, Figure 3.8 shows which questions are already tested empirically in literature. The arrows labelled with *L* refer to information flow based on the legitimacy theory. There is sufficient information available how the general public is putting pressures on investors and companies with supporting regulative ac-

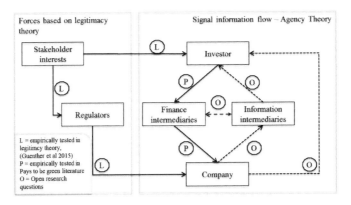

Figure 3.8: Research gaps in environmental disclosure agency setting

tions (Pattberg, 2012; Chithambo and Tauringana, 2014). The arrows labeled with *P* refer to money flows by investors and finance intermediaries who invest their money according to a green investment strategy. The financial success of this investment strategy is part of the pays-to-be-green research (Endrikat et al., 2014). Many ideas have been tested in the field of the pays-to-be-green research question, even though the lack of poor proxies remain (Arnold, 2011; Delmas et al., 2013; Greenwald, 2014; Lackmann, 2010; Lee, 2012; Mefford, 2011; Murguia and Lence, 2015; Oberndorfer et al., 2011; Robinson et al., 2011; Searcy and Elkhawas, 2012; Walther et al., 2009). The arrow *I* between finance and information intermediaries stands for the interpretation of the signal. Here, the amount and robustness of disclosure could be interpreted as a company that acts in a particularly environmentally friendly way and/or has incorporated climate change particularly well into its strategy.

The current reporting practice fails to provide sufficient information to judge the efficiency of companies. This problem of insufficient data, however, is ignored by the dominant research stream in the field comparing financial and carbon performance. This absence of critical feedback permits companies to claim green performance without the need to provide evidence. Therefore, it is recommended to focus more research on ways of identifying reports that aiming to achieve comparability and enable quantitative-focussed researchers and investors to identify high-quality reports. In addition, researchers should focus on exploring how currently available information can be used to judge how companies perform compared to scientific targets (Niehues and Dutzi, 2018).

Chapter 4: Qualitative study

A. *Qualitative research questions*

Having determined that the exploratory sequential design contains a qualitative research question, a quantitative hypotheses and a mixed-method research question (Creswell, 2014, p. 152), it is important to understand how these questions interact. The qualitative study has the purpose to understand current carbon accounting and data interpretation in terms of messages. The underlying research question is thus:

> How can carbon numbers be interpreted and what are the limitations in data quality?

The results of this qualitative question will be used to further specify the mixed-method research question and formulate the respective quantitative hypotheses. Figure 4.1 illustrates the interaction of the research questions. The first step shows the starting point of the mixed-method purpose statement. This statement defines the goals and questions for the qualitative study. The outcome of the study influences the mixed-method research question, which defines the hypotheses of the quantitative study. Looking at the benefits of the qualitative study, the purpose is to evaluate and refine the theoretical model, identify the relevant hypotheses for the quantitative study and collect, as a by-product, information about measures relevant to operationalizing the model. This process implies that the steps commonly used in quantitative research - problem formulation, development of a theoretical model, operationalization of the theoretical model - are already embedded in earlier steps, and that the quantitative study can directly begin with the data-collection plan, collection, and analysis. The interpretation of the result will already be part of the mixed-method interpretation as the interpretation should also include the respective qualitative findings (Kaya, 2009, p. 49). Translating this model of research questions into the narrative design of this document, Figure 4.2 shows the sequence used in this document. The bigger green circles indicate the mixed-methods, the yellow stands for qualitative parts, and the blue area stands for the quantitative part.

The PURPOSE of the qualitative research part is to explore how experts perceive the usability of carbon reports. Here, the focus of the expert interviews is the question of how to distinguish between high- and low-quality reports, and what information can be retrieved from high-quality reports. The

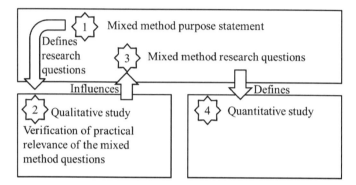

Figure 4.1: Development of the research questions.

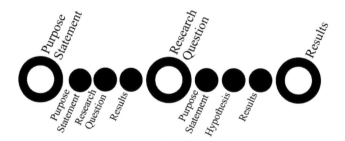

Figure 4.2: Narrative structure of the research.

research question is: if GHG disclosures are clustered in high-quality and low-quality reports, what would experts do to distinguish between carbon efficient companies and non-carbon efficient companies (Richards and Morse, 2013, pp. 120-128).

This general question implies several incidental findings. As experts define the quality step in which actions must be taken to identify reports that are comparable to one another, these findings must be interpreted for regulators and stakeholders to be defined as minimum requirements for GHG disclosure. At the same time, experts potentially contribute evidence of the things companies do to defeat efforts to compare companies. These actions can be interpreted as green-washing strategies. The third area of learning from these expert interviews is an important cross-check. Relying on the expert views, it should be possible to evaluate whether current quantitative mainstream research (so called *pays-to-be-green* research) really reflects the practical need of the stakeholders interested in environmental performance, or whether other research questions currently have greater relevance. These results are especially important for further discussions of the result with the ESG research community.

Another aim of this research is to ensure that the quantitative part of the mixed-method study has practical relevance, and that the findings can add value for regulators, reporting companies and ESG analysts.

The general setting of qualitative research typically consists of one or two OPEN RESEARCH QUESTIONS. These can potentially be specified in greater detail by asking more detailed sub-questions over the course of the interview. As qualitative research aims to understand the research area better, and not to generate empirically valid results, the thinking of participants should not be narrowed down in any particular direction. To keep the interviews exploratory, the starting question should be as broad as possible. Depending on the results of the first interviews, the broad question can also change over time to incorporate new knowledge generated during the interviews (Creswell, 2014, pp. 139-142). Therefore, the question of the interview was defined around how the expert would identify green companies, and what steps experts would recommend to companies seeking to be perceived as a green company compared to peer companies.

Based on this information need, experts should be able to analyze carbon reports in such a way that green companies are identified. Determining the group of relevant experts is crucial. One group of experts would definitely be the group of users of the disclosed information. As the theoretical model builds upon the agency theory, the respective persons could then be the principals who are supposed to interpret the signal. As this communication implies at least two parties every time, it would also be relevant to interview agents who voluntarily send the signal, as these agents are likely to know which signals they can easily copy and which signals should enable the principal to interpret the results.

As supported by the agency theory, the trustworthiness of disclosures can be increased by voluntary audit. These specialized auditors should thus also be able to judge the comparability of audited reports. As investors are often driven by financial motivations for wanting to invest in green products, there might also be a market for information intermediaries who interpret green signals. These specialized analysts should also be surveyed (Guenther et al., 2015).

However, the voluntary disclosure is also influenced by further actors that exert a strong influence. The interest groups which push for voluntary and mandatory disclosure are likely to include experts who can interpret the results of the requested disclosure (Guenther et al., 2015; Hahn et al., 2015; Bansal and Roth, 2000; Bansal, 2005). In summary, relevant experts can be found in

– Stakeholders which use the published data for analysis, e.g. investors, investor initiatives (most prominently CDP), consultancies in the financial sector, NGOs.
– Reporting companies and their respective service companies supporting the companies (e.g. consultancies, auditors & industry initiatives).
– Standard-setters that define how the data should be collected and published. This group contains governmental bodies, standard-setters for such general standards as the GHG Protocol (published by WRI & WBCSD), norming institutions (e.g. ISO) and all members of the respective advisory groups, as their knowledge is crucial in defining the respective norms.
– Researcher in the field of carbon disclosure.

There is no ranking applied with regard to which stakeholder group has the strongest influence. The ranking would also be quite challenging, as experts often represent more than one of the named fields. E.g. experts from (audit) companies are also involved in standard-setting processes, which are partially interlinked to regulators (Andrew and Cortese, 2013). Based on these different perspectives as to where potential experts can be found, it is now possible to define the population of experts.

B.   Expert population and sample generation

I.   Expert population

The approach taken to define the population and the sample taken targets the most effective way to reach the intended research outcome. Ideal is a representative random sample (quantitative approach) of the relevant population. In explorative research, however, the size of the relevant expert population is often unknown (and linked with the exact number and identity of the actual experts), and its attributes and distribution in a sample are also generally unknown (Flick, 2007; Meuser and Nagel, 1991; Lewis and Ritchie, 2006).

In the areas identified above in which experts could be found, the number of GHG-disclosing companies is endless, and the respective experts within the companies are often not disclosed externally. The same applies to audit companies, financial and ESG analysts and regulators and their corresponding experts. Additionally, regulators are also influenced by lobbyists with respective expertise. More transparent could be the standard-setters which often publish the list of individuals providing feedback for a standard. As standards often have quite a long history, the original experts, e.g. working on the GHGP 2004 (WBC, 2004), are not necessary the experts today. The same applies for academia. A potential solution here would be to sample based on structured literature reviews (Hahn and Kuehnen, 2013; Fifka, 2013; Aguinis and Glavas, 2012; Healy and Palepu, 2001, among various other).

Taking all the sampling issues together, it is questionable whether this manual definition of a base population and a respective random sampling would help arrive at more representative answers. First, the potential population is huge. The qualitative approach of open interviews cannot deliver representative answers anyway.

II.   Sampling strategy

At the same time, a completely random sample may have negative impacts on the heterogeneity of the experts sampled, and thus on the heterogeneity of views expressed (Flick, 2007). In such situations, it seems reasonable to rely on a theoretical sample (Lamnek, 2010; Glaser and Strauss, 1967) (Witzel, 1982, p. 35). The researcher decides what data are to be collected and where these data can be found. As the data are retrieved from expert interviews, the experts sampled will determine the respective outcome. Based on this sample, the expected outcome cannot be the quantification of a certain problem or its distribution, but it can be used to understand which problems exists in reality. As this study focuses on the nature of problems accounting companies face when calculating their group footprint, and how investors or other stakeholders can use these data, a theoretical sampling approach is reasonable.

Usually, theoretical sampling is done if the theoretical background is not empirically tested. Even though several elements in the field of carbon accounting have been tested, the results are often mixed. This could be the result of uncertainties for which empirical analysis do not account (Albertini, 2013; Hahn et al., 2015; Stechemesser and Guenther, 2012). Therefore, a theoretical sampling strategy is reasonable in this study. The sampling plan is developed based on this sampling strategy. A qualitative sampling plan requires knowledge about differences and potential working hypotheses upfront. This enables the researcher to predefine attributes or combinations of attributes to be incorporated in the sample. These indications are defined prior to field research and are incorporated in the sampling plan. The sampling can be done before the field research; in explorative research, the sample can also be defined during the field research. Due to the mismatch between the research fields identified in the literature review and the results of the qualitative study, it seems reasonable to sample during the study (Lamnek, 2010).

The interview sample selection can be performed in several ways. One option would be to aim for a random sample out of a given list of experts. As the experts come from heterogeneous groups, a separate random sampling should be done for each group. For experts from reporting companies, this would imply generating a list per expert group based on the area in which the respective experts are found. E.g. there could be a list of companies that provide high-quality reports as measured, for instance, based on the top quartile of the CDP transparency score. Relevant reporting experts and auditors could then be sampled from this list of companies. The same logic would apply to the group of standard-setters. First, a list of relevant standard-setter initiatives and standards should be compiled and the respective authors of standards identified thereafter. This could be based e.g. on the GHGP lists of authors and persons providing feedback. These lists could be used as a sample base. The group of investors could be covered by contacting the CDP and other dominant investor initiatives and sampling investors who signed the CDP. Alternatively, the researcher could gather a list of consultancies with a focus on green investments. Looking at the research purpose to identify best practices of cherry companies it is likely that the result of those interviews would be BIASED by the way of how the experts were gathered. E.g. reporting companies which scored especially well at the CDP are likely to propose the CDP score as good measure to identify cherry companies.

A second option is to look for an expert community that convenes the different experts, group them by expert group and sample from this community randomly. Potential groups might be the CDP conferences or respective CDSB members.

A third option is to sample based on recommendations made by the experts sampled. Here, a known expert needs to be identified to recommend someone in the field of data quality assurance. Based on a snowball system, each interviewed person would be asked to recommend further experts personally. Additionally, each expert would be surveyed about the areas in which he or she would search for experts. This question can be used to verify whether the sample used is valid for the research purpose involved.

OPTION ONE of the sampling strategy was discarded for several reasons. First of all, the usage of a certain list to find experts already constitutes a sample. As the full list of reporting companies, auditors and initiatives is unknown, this sample is not as random as it looks. At the same time, the risk of getting somebody with a relatively low expertise is quite high, as it is not clear whether the expert selected is new to the field or has changed jobs, meaning the information obtained would be outdated, or was included on this list with relatively little effort, e.g. as some specialist person on his or her team did the work, and the person on the list merely represents the company/initiative. At the same time, the size list of potential experts is very high. The relatively high number of potential experts is illustrated in the fact that more than 3,000 companies report data in the CDP. Even if the list of companies is reduced to the leadership group, the environmental reporting experts from these companies would still have to be identified. The standard section is no easier to assess. First, there must be an assessment of which standards are relevant, and second, each standard will have an unknown number of persons involved. The GHGP lists more than 500 contributors. According to the CDP, there are more than 200 legislative actions around the world in the field of carbon accounting. In most cases the relevant experts who develop those legislative actions are not publicly announced. Investors typically also do not disclose their environmental experts externally. As audit and other certification companies have identified environmental disclosure as a growing business market, multiple companies claim to have the relevant expertise in this field. While it is quite easy to identify those proposed experts it is rather hard to validate their expertise. As experts are sampled randomly from the respective list, the expected response rate is poor which biases the results further.

One argument for OPTION TWO is the strong influence on current methods in carbon accounting and its future development from cross-functional initiatives such as the CDSB. At the same time, the CDSB is already distributed across the various networks of experts (CDS, 2015a). This approach has also some weaknesses, however. First of all, one problem for this study is the fact that the seniority of the CDSB Board members (invited by the World Economic Forum) is typcially associated with a low response rate (Bednar and Westphal, 2007). This issue could be partially mitigated by inviting the CDSB Advisory Committee and technical working group members. The second argument against this sample approach involves the ability to receive the information requested. Sampled experts from the CDSB will potentially provide biased answers on the limitation of published CO_2 emissions, as the CDSB was founded to make reporting companies comparable. The bias could thus be either that the members might have an interest in not disclosing expert information on the source of uncertainty, or that the experts might not be aware of all of the issues involved. Another bias is the potential similarity of the answers: as these experts discuss the problem of uncertainty of carbon emissions amongst themselves, the views of the experts could become homogeneous over time.

OPTION THREE is to ask one expert to identify potential experts outside of his or her own organization, and to ask these experts for feedback on which experts they would recommend and where they would look for experts. This snowball sampling system is very common in qualitative research settings (Watkins and Gioia, 2015, p. 62). The target behind the approach of asking interviewees to introduce the researcher to other experts is to enlarge the sample over time. The aim of asking where experts can be found is to validate, following the study, whether the sample covers the respective areas, or if the sample needs to be enhanced (Lamnek, 2010). The success of this sampling strategy depends on the first expert and his or her respective network among carbon experts. At the same time, it is crucial that the experts agree to the research needs of this study, so that they invest their time and recommend the research further to other persons. Potential issues could be that the experts chosen represent only a similar viewpoint (e.g. all having a similar background) or do not have the respective expertise. Another limiting factor could be that these experts might want to influence the outcome in a certain direction (Struebing, 2004, 29-33). In light of the arguments presented, sampling strategy three was chosen.

III.   Sample description

Following the approach of the SAMPLING STRATEGY three described above, 15 experts were contacted; 11 of these voluntarily gave interviews. In four cases, a face-to-face interview was possible; the others were interviewed over the telephone. As a heterogeneous sample was intended, the sample is now described based on these different elements:

As shown in Table 4.1 the heterogeneity in work areas implies researcher (1), financial-information intermediaries (2), NGOs (4 experts work for environmental reporting NGOs including the CDP), reporting experts from companies with an accounting function (1) and a CSR/communication function (1), and two auditors (1) with a focus on sustainability, one auditor from a big four company and one auditor from a non-big four auditor.

Particularly interesting is the membership in different expert communities on the part of the representative sample and the respective interaction with other expert groups. The experts in the sample include members of the CDP (as full-time profession), accredited governmental experts for carbon regulation in Germany and France, other expert groups working in close cooperation with regulators, e.g. ICLEI network for cities or UN green freight action plan, members in the expert groups of harmonized reporting in the transport area (GLEC), members of the sustainability auditing standard-setter group for the European Union and for Germany, and several environmental industry groups. These imply the relevant United Nations-sponsored industry platform, industry-driver expert groups with national, regional and global scope, and global standard-setters with influence over the Paris agreement. Members of academic research groups for green reporting with national influence, e.g. on regulation for sustainability reporting and implications of integrated reporting in three different countries, were also interviewed. From a rating perspective, experts and rating partners from four different initiatives with global scope were part of the sample. Every expert has published at least online guidelines with regard to voluntary ESG disclosure as part of his or her main job. About half of the experts also published in national and international academic journals. This is especially interesting, as only one interview partner holds his main employment contract in academia.

Area of expertise	Interview			Thereof involved in	
	Person	Phone	Rejected	Standards	Regulation
Company	2	0	0	2	0
Auditors	0	2	2	1	1
NGO Investor	0	2	0	2	0
NGO Non-investor	1	1	0	2	1
Investor	1	1	2	1	1
Academic	0	1	0	0	0
Total	4	7	4	8	3

Table 4.1: Overview of interviewed experts (Niehues and Dutzi, 2016).

Several participants' names could be found listed as contributors to standards. This involves all four versions of the GRI standard, the GHG Scopes 1 & 2 protocol, the GHG Scope 3 protocol, the GHGP product standard, DIN and ISO standard 16258, the ISO standard for calculation of environmental costs, the GLEC framework for globally aligned logistics reports, the IDW sustainability audit standard, and many more. An overview of the full list of standard and expert communities would enable the reader to identify the interviewed experts. This is not intended as it was indicated during the interview that the results would be treated anonymously. For some of the named standards the interviewed person was in charge of publishing the respective standard. The hierarchy level of the persons interviewed ranged from non-leading professional level to managing director. Interviewed persons include male and female interview partners. The age of the interviewees ranged from late twenties to sixties. The fact that within this sample only a limited number of people have significant influence on various reporting standards shows the validity of the criticism that a limited number of players influence reporting and even legislation (Andrew and Cortese, 2011b,a). At the same time, it shows that the experts sampled do indeed belong to the required *functional elite* (Meuser and Nagel, 1994, p. 182) and should therefore have the requisite expertise for the underlying research. In terms of regional geographic setting, nine experts hold a global role, and only one expert focuses on national issues. The interviewed experts had work experience on all continents during their careers. In terms of current working location, no country was represented more than four times (Germany). The other countries in which the interviewed persons had their main office are twice in the

USA, the Netherlands and Switzerland. One participant came from the UK. This over-representation of the Western world also reflects the dominance seen in other environmental networks e.g. CDSB, CDP, WRI, WWF etc (Andrew and Cortese, 2013). The level of education ranges from Bachelor's degree to PhD & Professor. The different educational backgrounds of the interview partners range from history and data science to various degrees in business (e.g. accounting, environmental accounting, marketing, business-IT) and environmental engineering. This information was not fully collected during the interview. Instead, this information was captured based on their self-published information in business networks (mainly Linkedin.com), as part of the author information in publications, or online information of the organizations with which they are currently employed.

As the small size of of the expert sample does not permit any empirically relevant result, the heterogeneity of the sample suggests that the answers will permit the reader to understand various viewpoints in respect of comparability of figures. One potential limitation of the sample is its focus on experts in the ESG field and the fact that it does not also include analysts without ESG focus which somehow accounts for carbon as a risk factor. This limitation is due to the focus of the research question on the things that can already be done with carbon figures. Therefore, the focus is to understand the ability to identify the best companies. This is potentially most easily done by experts focusing on green disclosure. At the same time, it should be acknowledged that not all reporting companies and analysts do have this highly specialized expertise available. Due to the sample generation and the snowball system of the experts, the sample is biased based on the experts interviewed (Kuckartz, 2014, p. 84). This bias increases due to the limitation in the response rate, even though the responses rate of over 75% is very high. A limitation in the responses concerns the actual usage of information by investors, as use of this information is typically proprietary information. Even though one financial analyst showed an sample study of a client, this study was shown as restricted material which cannot be quoted within this study.

C. Interview guideline

The interviews were conducted based on a semi-structured interview guideline. The interview was recorded, unless participants preferred an interview without recording, which was the case four times. After & during the interview, notes were taken.

The interview is a problem-centric interview (Lamnek, 2010) in which the interviewing researcher has already developed a theoretical concept for an identified issue, and the interview is meant to focus on the respective research question (understanding the reasons for uncertainty in carbon numbers and searching for KPIs used in practice to identify carbon-efficient companies). This kind of interview uses open questions, encouraging the interviewed persons to explain as much as possible.

The interview is clustered into four or five phases, including an introduction; general remarks by the interviewee; specific questions to ensure that the given content is interpreted correctly, e.g. by rephrasing how the interviewer understood the points, asking for clarification or highlighting conflicting statements made by the interviewee; ad-hoc questions about issues not raised previously; and, finally, potentially a standardized short questionnaire. This interview can be supported by a short questionnaire, an interview frame, an video/audio taping and a postscript in which the interviewing person provides context [Flick, 2007, Lamnek, 2010, pp. 212-219][1]. The introduction is given by the interviewer, explaining the research aim (finding ways to compare companies based on carbon data) and presenting how the researcher arrived at this research question based on his own experience. This helps the expert consider the interviewer as an expert at eye level (Creswell, 2014, pp. 139-142). During the introduction, the interviewer also highlights why the expert was nominated; this encourages the expert to provide more information. Afterwards, the expert is asked about his or her professional background to ensure expert validity and to break the ice for the main research question. As the core research question is very detailed, the interview opens with two guiding questions (Creswell, 2014, pp. 139-142). These are:

Why do companies voluntarily report GHG emissions? The background for this question is to validate the theoretical model and to have the expert focus on voluntary reports.

How can an external person identify whether a company is efficient or non-carbon-efficient?

As for the first question, plenty of arguments are available in the literature. It is expected that the experts will have no difficulty responding here. The second question, however, is very critically discussed in the literature, and with no real answer. Therefore, the experts' response here are expected to be evasive. Accordingly, some associated sub-questions have been developed to

---

1 The interview guideline is shown in attachment 6.2, the audio files, and postscripts for all interviews are part of the digital attachment.

facilitate the effort to narrow down the answer. For direct answers tending in the direction of this is not possible, the sub-question is, why this should not be possible or what are the prerequisite conditions for KPIs are to enable comparability. Answers should then be rephrased to ensure a common understanding of limitations of CO_2 figures, and the KPIs the experts have in mind should be noted. For answers which can be summarized as difficult to answer, the associated sub-question is to ask how would you start or what would you recommend to look at if you are tasked to do so. Over the course of the interview, it should be ensured that the expert is providing his or her information on the following elements identified in the literature as limiting factors for comparisons of carbon inventory (Kromrey, 2010, p. 364):

- *How important is voluntarily reported information for comparisons of CO_2 efficiency?* This question aims to understand if the disclosed information is seen as positive signal which enables stakeholders to differentiate between cherries and lemons.
- *How to identify CO_2 efficient companies? What are determinants of efficient companies?* Focus of this question is to identify operationalization factors on a company level.
- *What are the determinants of a good report?* Focus of this question is to identify operationalization factors on a report level.
- *Which factors limit the comparability and accuracy of CO_2 figures?* Within this question, it is important to identify which of the factors, namely complete & identical scope, reporting standard, existence of an audit and its respective level of assurance and the audit provider, the calculation method, the emission factors and the data used. Here it is interesting to see which of these factors is/are proactively named by the expert, and which are confirmed after being asked. Aim here is to understand if the research streams focus on relevant parts of if the experts have different perception of the carbon disclosure practise.
- *Could you quantify the amount of uncertainty in the data provided?* This question aims to understand in how far published KPIs are valuable for comparison or if the underlying uncertainty outweighs the provided information.
- *Which of the factors named are also used by investors to analyze the efficiency of the company?* Ideally the feedback can be used in the quantitative study.

After this first content of the interview is complete, the next section aims to proof the sample and find an interview partner for the next interviews (snowball sampling). The first question asks what areas the expert would look to to find relevant interview partner. E.g. if the expert is an auditor, the question would be *where do you think experts for these questions could be found, besides at audit companies?* The aim of this question is to evaluate whether all kinds of different views are represented in the theoretical sampling (Creswell, 2014, Watkins, 2015, p. 62). The next question is the snowball sampling idea to ask whom the expert would recommend to request for an interview.

D.   Evaluation of feedback and its limitations

As the interview guideline also provides the sub-categories for reviewing the results, the working steps of transcribing the interview and coding the interview phrases into the category system could be combined in a single step (Gläser and Laudel, 2010, p. 200). The category system antedated the inter-view guideline and was plausibility checked in an upfront expert interview to ensure that the category system is as complete as possible. One argument for this approach is that the experts are reporting on facts and/or the approaches they would take to analyze data. Therefore, the ability to capture emotions on this topic is not as important as in other social science questions for which qualitative interviews are used more often. Additionally, the application of the transcription is not used consistently as it is also recommended to record information provided outside the recording time (Lamnek, 2010). One limi-tation of the interview and its documentation is, that four experts declined to have audio recordings made of their responses. In this cases the notes constitute the main information were confirmed by the interviewed experts. Another limitation of the interviews were that the interviewer includes his own work experience and expertise in the interviews. This reduces potentially the validity, as the reproducibility with a triangulation of the interviewer would be different. However, this limitation in qualitative research is often seen as acceptable as expertise of the interviewer leads often to increased information during the interview as the interviewed person is more likely to provide insider information (Creswell, 2011, pp. 232-235, Silverman, 2006).

Once the categorized information (sub-category level) has been clustered, the results are structured into five areas, namely findings consistent with the criticism found in the literature, findings that are the topic of controversial debates among experts, findings different to what the literature suggests, literature findings viewed as irrelevant by the experts, focus areas of the experts that are currently the in focus of the research agenda.

E.   Qualitative results

One outcome of the interviews is that all experts find it difficult to compare carbon efficiency between companies. Given the set of information currently externally available, no expert was willing to say he could judge carbon efficiency based on the published results. Several experts indicated that comparisons of efficiency between companies can be done only for similar products at a product level. This product calculation shall be calculated consistently with group reporting. It is mandatory that these footprints contain scopes 1-3, and that the boundaries are set similarly/identically.

The reasons for carbon reporting are explained as a combination of stakeholder pressure and the aim to present the own company as environmentally friendly, or at least to limit the negative perception of stakeholders. All experts agree that there is a significant number of published carbon reports have wrong/incomplete/inconsistent data. This is mainly not perceived as fraud but rather as a consequence of missing competencies in the reporting companies or the limitations of available data. Especially experts who working empirically with carbon numbers (interviews 1, 5, 7, and 11) highlight that the figures submitted regularly lack a plausibility check, as they are over- / understated by e.g. a factor of 1,000 as units of measure are used incorrectly.

Many experts (interviews 1, 2, 3, 5) highlight that reporting companies disclose similiar information through different channels with inconsistent data. Here, the experts would like to see stronger regulation and stakeholder feedback in the form of refusal to accept these diverging numbers. Looking at the comparability of figures, one outcome of the expert interviews is that a significance gap exists between the proxies which should be used to determine whether a company is carbon-efficient, and the proxies currently used by stakeholders, e.g. investors and researchers. When asked on what level companies are being questioned by investors and regulators, usually scope 1 or scopes 1 & 2 emissions divided by either revenue or number of employees is used (interview 1, 6, 7, and 9). This stands in direct conflict with the main expert opinion that all carbon-intensity KPIs which do not include scope 3 emissions are useless, as they measure only the outsourcing activity.

Generally, experts are more interested in the trending, target-setting and target achievement of carbon information than in comparisons with other companies. For this purpose, investors aim to get reliable timelines, ideally in comparable databases, as the CDP provides. In terms of timelines, experts point to their experience that a company's carbon disclosure follows a maturity curve, so that first reporting is almost always incorrect and data quality improves over time. This could limit the assessment of efficiency trends for companies with a short reporting history. The financial intermediate interviewed (interviews 1, 5, 7) solved this issue by estimating CO_2 numbers for companies. In addition, a reliability score for the reported and estimated number is calculated. For analysis, only carbon numbers above a certain reliability threshold are used. Whether and, if so, to what extent the number of years of reporting forms part of the score was not disclosed during the interview.

In terms of boundaries, some experts highlighted that it is important to be transparent about the extent to which a company plays a role in being the solution to fight climate change, or being the cause. This statement also refers to the product level. The underlying question is whether the reporting company produces e.g. clean energy products or products that are dependent on fossil fuels. In terms of required accuracy, the experts have different approaches about what has to be done in order to compare companies. All experts see the requirement as one of comparing only complete footprints with similar guidance on how the emissions are defined. With the exception of one expert (interview 7), all experts see the need for external assurance. Most experts (interviews 1, 4, 5, 8, 9), even one of the two experts from audit companies, see limited assurance as sufficient.

The minimum requirements for the data to be used in a comparison vary significantly. Experts with an analytical background of looking to published numbers from an external view (auditors obviously have access to internal data) have rather broad minimum standards that are often captured in large databases. Extreme opinions are found here major scopes 1 & 2 emissions, at least 70% are sufficient. Others require a complete footprint with full financial boundaries, a reasonably small standard error over time and reporting based on a known standard.

By contrast, experts who are involved in company reporting (or discussing these activities in standard-setter boards) look into the data with much greater detail. To begin with, these experts are take a detailed look at scope 3, not only as an element but on the level of determining which are the relevant scope 3 categories in the industry, what the data sources for scope 3 are and which methods are used to calculate emissions as a way to compare like for like. The next element in which the experts look in greater detail is the relevant efficiency KPI. While empirical experts use revenue or other generally available KPIs, reporting experts look into the product level of companies and sector-specific KPIs. Generally, experts see the application of certain standards as linked with accuracy and completeness of the footprint. Some experts can judge the quality of a report based on the reporting & audit standard applied.

In terms of the usability of the carbon reports, there is a shared view that the GHG inventory is one KPI that can be used to tell a broader environmental story about the reporting company (most prominently highlighted in interview 3, also supported in interviews 1, 2, 4, 8). The company is still requested to actively report on its climate-change agenda, and to link this information with the trends of the inventory in the respective categories. Some investors use the amount of disclosure as one indicator for evaluating a company's climate-change strategy.

In terms investment strategies, two main strategies were highlighted. The reactive strategy is used to identify hot spots within an investment portfolio. During investment discussions, investors can either focus on companies from these hot spots or divest their holdings in these companies (Allianz S.E., 2015; Bergius, 2015; CDP, 2015c; The, 2016). The financial experts interviewed actively state that the investment decision is not taken purely on the basis of carbon-efficiency KPIs.

Stock selection based on CO_2 efficiency KPIs would be really stupid. I cannot imagine that any institutional investor would do this. -Interview 1

This expert statement is contradicted by ETFs, which actually use carbon-efficiency KPIs to set up their portfolio.

The proactive strategy identifies only companies that focus their value chain on products that actively reduce climate-change effects. In both cases, the carbon numbers are only one piece of information which is supported by a qualitative interpretation afterwards. Only three experts (interviews 3, 4, 5) were willing to actually make a real comparison between companies based on their efficiency. However, these experts required very detailed data on standards, audit levels, asset composition and product portfolio which is currently a detailed qualitative assessment. Even though only a few companies, 224 globally as of 30 March 2017 (sig, 2017), set themselves a scientific target, almost all experts highlight the need for scientific targets. Additionally, experts highlight that these targets needs to be published externally, including honest reporting of the extent to which the company is on track (interviews 3, 10, and 11). Explaining why a company is currently not on track, and what actions are planned to change this, is seen as a signal of honesty. This is a major disconnect to the current behavior of the reporting companies, the investor community, and the existing regulations, which are mainly disconnected from the scientific target-setting process. Connected with this scientific target is also the method of disclosing information. The experts who made any comments about managing carbon inventory highlighted the need for absolute reduction targets embedded in the company growth plans. This means that each company must improve its relative KPIs and while at the same time reducing its total carbon footprint across all scopes.

F. Interpretation of results with respect to mixed-method purpose statement and research gaps

Looking at the outcome of the interviews, it is obvious that absolute carbon numbers involve immense internal complexity. This limits the comparability of companies based on their inventory. Additionally, the output of companies is very different across sectors, and this limits the ability to find common intensity KPIs. Currently no cross-sector KPIs are available to enable stakeholders to compare companies based on carbon intensity. This is the main reason for the existence of very different methods of carbon disclosure. However, even though carbon disclosures cannot facilitate direct comparisons among companies, the experts did clearly articulate how to differentiate between a CO_2 disclosure that represents a true and fair view of a company and a CO_2 disclosure that does not aim to provide any transparency of the company. The most crucial element was the complete and consistent reporting of emissions over time. The feedback that companies regularly omit material areas of their carbon footprint was frequently stressed by all kinds of experts. With the exception of one expert (interview 9), all experts considered completeness to comprise all relevant emissions within the sector for scopes 1-3. Only one expert considered scopes 1&2 as sufficient (interview 9); however, several experts (interview 1, 5, 6, 7, and 9) use only scopes 1&2, due to missing scope 3 data. Hence, any data analyses on carbon trending/carbon reporting must ensure that the companies are filtered to companies which report, alongside scopes 1 & 2, all material categories for scope 3. Another reporting element is the application of an acceptable reporting standard and external verification with (limited) assurance of the reported numbers. Following these elements, experts (interviews 1, 2, 3, and 5) interpret the data submission as sufficient to their reporting needs. Even though the disclosure does not necessary correspond with carbon strategy, but rather is *only one hygienic element in a carbon strategy* (interview 10), it is seen as *an indicator for the existence for a climate change strategy* (interview 1). The high-quality carbon report is additional evidence for stakeholders to verify that the company's CO_2 emissions are trending in the direction highlighted in the sector guidance for the company. This trending of emissions should cover an absolute trend as well as a relative-intensity trend based on e.g. revenue. The additional evidence must be stressed, as almost all stakeholders expect the company to pro-actively explain its own climate change story embedded in the strategy. The carbon disclosure is only the evidence that this story is implemented in the company on large scale.

Comparing the outcome of the expert interviews with the literature review, relevant research gaps can be identified. The first research gap concerns the data quality of CSR reports in general and of CO_2 disclosures in particular. Currently, academia fails to qualitatively assess what disclosures should entail to ensure that the companies can differentiate themselves based on these disclosures. These differentiating elements should be quantitatively verified. Additionally, academia could help stakeholders identify green-washing companies. This identification should ideally be cast such that the results can be applied in quantitative ways. The interview results could be used to develop minimum reporting requirements. These results should be incorporated before using any carbon inventory number in a quantitative setting. CO_2 disclosure should not be accepted unless:

– Companies report identical figures over all distribution channels.
– The report contains scopes 1-3. Scope 3 implies all material categories.
– The disclosure is consistent over time.
– The disclosure is externally verified with at least limited assurance.
– CO_2 reduction is embedded in the overall strategy of the company.

In an agency setting, the failure to meet those minimum requirements could be interpreted as a green washing signal.

Another research gap is the identification of relevant KPIs for each sector that reflect CO_2 intensity. CO_2 scope 1 divided by revenue is regularly used as a proxy in the investment community; this is not sufficient as this KPI can be easily manipulated by outsourcing emission intensive activities which reduces the scope 1 emissions without having any environmental impact. Developing sector-specific KPIs could also help identify sector classifications that reflect company similiarities in the field of CO_2 reporting. For experts, the empirical studies that dominate the the publication landscape for the pays-to-be-green area have almost no impact. This owes to the poor quality of the proxies used to identify green companies, limiting any interpretations (Niehues and Dutzi, 2016). The areas of audit gap and data accuracy seem to be not directly in the focus of most experts. Quotes from experts include:

The level of assurance is not relevant for the assessment of companies. - Interview 5

I am not aware that any stakeholder values the fact that we undertook the audit on reasonable assurance level. - Interview 2

The audit and its level of assurance is one among several elements to assess the reliability of the report. - Interview 1

Even among auditors, the importance of data accuracy is mixed and range from

Many of our customers are lacking the competence to communicate all their great activities to stakeholders. We guide them on how to collect which data and how to communicate this. An assurance statement is already a great success. In this setting the extra effort of a reasonable assurance has only limited benefit. - Interview 8

to:

I am regularly surprised, sometimes even shocked, about the contents in ISAE 3000 assurance statements provided by competitors. This is especially valid for smaller firms, certification bodies and boutique CSR consultancies. Our statements become copied and applied in a wrong context, e.g. when a non-audit firm refers to quality standards and requirements of professional accountants, or cases when listed audit procedures simply don't make sense in relation to the subject matter of the engagement. The lack of quality sometimes even becomes obvious in the report-level disclosure, where I can spot errors by mental arithmetic ratios. - Interview 11

As carbon submission is seen as only one element for the analyses, and the aim is to reduce emissions down to 0, uncertainties within the numbers are accepted as long as reports enable companies to identify their hot spots. It is expected that these hot spots must be replaced by completely different technology/processes. The question of whether the corresponding carbon number represents a CO_2 inventory of 80, 100 or 120 is irrelevant.

The total number is not so relevant. The core question is, how quickly is the company able to reduce its emissions to zero or below. That is the question which enables me as stakeholder to judge if the company is part of the problem or part of the solution. ... of course emissions include upstream and downstream emissions. ... The absolute number and its audit level is here of lower priority. - Interview 10

The experts seem to place high trust in standard-setter and ISO norms, as the calculation method used was a point very seldom raised. And even if actively questioned, most experts highlighted that all methods accepted by (any) ISO norms, industry frameworks or standards are acceptable.

This high trust in standards and industry norms opens research into the practical implications of standards, the differences involved, and, potentially, the grading schemes used. This is especially true, as where the implications of using certain standards and calculation methods are concerned, there is an information gap between reporting experts and the stakeholders using the data (Andrew and Cortese, 2011a, 2013, p. 400).

The element experts spontaneously identified most often consisted of the challenges resulting from the two-degree target. Because the two-degree target is seen as the core expectation that society has of the companies, disclosure should aim to guide stakeholders to interpret the extent to which companies are supporting this target. In this area, several research gaps can be identified, starting with the question of how stakeholders can analyze whether companies are meeting the two-degree scenario. Currently, stakeholders are dependent on companies' internal scientific targets. Therefore, the academic community needs to support the breakdown of the macroeconomic data to include the implications of the reporting company for the two-degree scenario, and translate these implications to KPIs that can be externally calculated. KPIs that can be used to track a company's status towards the two-degree target are currently missing (Walls et al., 2011). The need for this ability to assess progress on the path to the two-degree target opens up research fields on the implications for the two-degree target. Based on these findings, a mixed-method purpose statement can be developed that forms the basis for quantitative research questions (Creswell, 2014).

Chapter 5: Quantitative study

A.   *Carbon disclosure as signal for green performance*

I.   Purpose statement

Taking the different disclosure strategies as shown in Figure 3.7 into account, the experts highlighted in the interviews the difficulty to identify truly green signals. The different options of signaling between investors, stakeholders and companies as shwon in Figure 3.7 can be reduced to truly green investors and stakeholders (option 3) and companies which are sending truly green signals (option 3) and false/green washing signals (option 4). Figure 5.1 shows the reduced options. Based on this reduced disclosure strategy options, the theoretical model of agency signaling as shown in Figure 3.6 can be used to map the research questions as shown in Figure 3.8. Building upon the feedback of the interviews and the construct shown in Figure 3.8 the purpose of this quantitative research is to develop a measure that reflects an assessment of the quality of the disclosure and test whether the quality of the disclosure is impacted by the underlaying green performance. The underlying green performance should ideally be measured against the two-degree target as this was identified as further missing link. Additionally, it should be investigated if information intermediaries support this messaging. These aims can be transferred to hypotheses.

II.   Main hypotheses

Assuming that companies disclose CO_2 information to stakeholders to prove that they are acting in an environmentally friendly way and taking respective actions to reduce their carbon inventory over time, emissions ought to decrease over time. The first hypothesis thus tests the general link between carbon disclosure and carbon performance.

> Hypothesis 1:
> Companies with positive environmental performance disclose better comparable information than companies with less good performance.

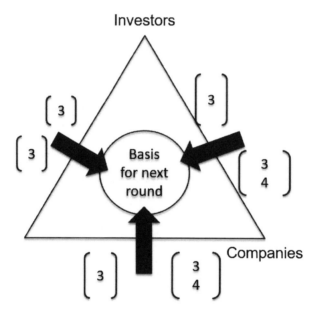

Figure 5.1: Options for companies facing green requirements.

As this relationship provides only a general trend between carbon disclosure and carbon performance, it is relevant to understand whether stakeholders can judge, based on the company's disclosure practice, if a company is meeting societal expectations that it sufficiently reduces its CO_2 emissions. After Paris 2015 and the respective IPCC report, this stakeholder expectation can be defined as the two-degree target. The question thus becomes whether the reporting company's current CO_2 emissions trend is compliant with scientific target-setting pathways. This question is formulated into the following

Hypothesis 2:
The majority of companies that send a positive CO_2 disclosure signal meet the scientific two-degree target.

Within the signaling discussion, the question arises as to how the agent can ensure that its message is trustworthy. One typical way to ensure this is to find an information intermediary that will ensure that the content of the message is correct. The reputation of the intermediary guarantees that the intermediary has checked this information. In the disclosure field, an audit statement from a known auditor ensures this (Erlei et al., 2007). Therefore, Hypothesis 1 can be rephrased from more information to audited information.

Hypothesis 3:
Companies with positive environmental performance use voluntary audit services more often than companies with a poor performance do.

As indicated in Figure 3.8 several effects what influences voluntary disclosure have been already tested empirically. Therefore the robustness of the quantitative model can be increased by taking known stakeholder effects into account. This can be also seen as a quality check of the operationalization of the model. For this reason control hypotheses were developed.

III.   Control hypotheses to proof mediator variables

As Figure 5.2 shows, there are also other reasons for companies to improve their voluntary GHG disclosure. This causal effect of stakeholder pressure needs to be adjusted in an empirical setting. For this reason, conditional hypotheses (C-Hypotheses) test assumptions about mediator effects that influence voluntary CO_2 disclosure. C-Hypothesis 4 verifies the effect of stakeholder salience / pressure on companies to disclose more information. This principle was already tested empirically on the basis of the CDP disclosure score in a panel study from 2008 to 2011 (Guenther et al., 2015).

C-Hypothesis 4:
Companies voluntarily disclose information as a result of stakeholder pressure. The amount & quality of environmental disclosure is a function of stakeholder salience.

As disclosure costs depend on the size of the company, and as big companies are under greater focus by stakeholder interest groups than small companies, company size should also be used as a moderator variable.

C-Hypothesis 5:
Big companies disclose more information than small companies.

Discussions of the effect of financial strength can be controversial. As this study uses agency signaling theory as its theoretic underpinning, and accepts the positive causal relationship between financial performance and environmental performance, positive financial performance should be expected as an effect of green action. Hence, this research paper argues that financial performance should not be used as a mediator/moderator variable.

Another mediating effect is expected from the sector in which the companies do business. This mediating effect can be explained by agency theory, as the signal per sector might be adjusted for sector-relevant information. The principal (stakeholder) and their interests are likely to differ from one sector to the next as well.

From a legitimacy standpoint, stakeholder salience and pressure varies by sector. Looking at the isomorphic forces of institutional theory, it is likely that peer companies will vary by sector as well as due to other effects.

From an empirical standpoint, Patten showed in a meta analysis that company size and industry sector are relevant moderator effects for empirical tests within a particular social and environmental research domain (Patten, 2002, 2015). For this reason, the C-Hypothesis examines the degree to which this assessment can be verified.

C-Hypothesis 6:
The amount of disclosure is a function of the sector of the company.

Due to the strong empirical evidence that sector classification is relevant, models showing no significant effect on Hypothesis 6 will be discarded as not plausible.

IV.   Embedding the hypotheses into the theoretical model

Figure 5.2 shows how the hypotheses of the theoretical model of agency signaling are connected. The arrow between the disclosing company and the investor is connected to the first and second hypothesis as in a signaling model, cherry companies have the incentive to disclose comparable information. The arrows between the reporting company and the information intermediate to the investor can be connected to the third hypothesis. From a theoretic point of view the auditors reputation would be used to confirm that the sending company is truly green. Green washing companies could however also use the complexity of carbon emission reporting as argument to strive for the same confirmation by the information intermediate as the truly green company. The control hypotheses four, five and six are related labelled with L.

B. Operationalization

I. Reporting companies sample

The agency model shown in Figure 5.2 reviews only companies which strive to be identified as green companies. This model implies that the scoping of relevant companies is already part of the operationalization as companies which aim to send a positive message need to be identified. Therefore it is reasonable to identify a set of companies disclosing voluntary. Chapter E. identified here the CDP as area of information which is often used in research. The CDP data set is the largest data base of voluntary disclosed CO_2 information by companies and is used also by investors which fits to the agency model used. The information submitted to the CDP reflects therefore directly the message companies want to send to their investors. This direct access to the answers from companies is a big advantage to environmental scores as provided e.g. by KLD, as scores already include the interpretation of information intermediates. Other databases from e.g. Southpole use data from various sources including mandatory reporting schemes and self estimation of data which increases the number of companies in scope, however, these companies do not necessarily fit to the agency model as the companies do not necessarily strive to send a green message. The structured CDP questionnaire enables further operationalization to gain comparable information. This reduces any bias from e.g. interpreting CSR reports which often include also carbon emissions. To use as many reporting companies as possible all companies reporting voluntarily to the CDP were included (excluding the sectors banking and insurance). As reporting years the years 2010 till 2015 were used as 2010 was the year when the first questionnaire was send to

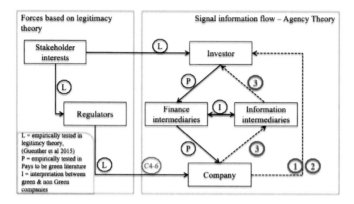

Figure 5.2: Interdependencies between the theoretical model and the hypotheses.

reporting companies and 2015 was the latest available data. From this data set the effects for the six hypothesis have to be operationalized.

II.   Environmental performance

Environmental performance can be measured in several ways. A common way to measure environmental performance is to enlist specific environmental ratings that rate companies based on their environmental performance. These ratings, however, generally only measure scopes 1 & 2, which can be explained by the poor data availability of scope 3, and more critically also take the availability of transparency of the information into account in its grading system. From a stakeholder perspective, it is fair to downgrade companies that are unwilling to provide transparency for this information. However, to statistically test whether disclosure practice influences environmental disclosure, a link in the grading scheme between disclosure practice and environmental practice makes the test redundant. As an example for the link between performance grading and disclosure practice, the CDP performance band grades the availability of a third-party verification so high that an A-performance score is nearly impossible to reach without external verification (source: Interview partner CDP). One approach to measure the trend in CO_2 emissions is to operationalize the efforts companies take to manage their carbon inventory. The CDP questions about downstream reduction initiatives and general emission-reduction initiatives can be transferred to binary variables representing these efforts. The same applies if companies set themselves absolute and/or relative targets to reduce CO_2. Alternatively, CO_2 performance can be measured by the absolute and relative trend of emissions. This is going to be done in three different ways. First, the data of the absolute scopes 1 & 2 trend is used. Here, the CDP questionnaire explicitly asks how absolute emissions for scopes 1 & 2 have changed compared year-over-year (CDP, 2015i, pp. 11-16). As Figure 5.3 shows, these self-reported changes in emissions are highly affected by 27 outliers, which might often not represent an operational change in emissions. Therefore, outlier values are replaced with the boundary value, which is calculated by multiplying the standard deviation by 1.5.

	Min.	1st Qu.	Median	Mean	3rd Qu.	Max.
Before	-100.00	0.00	0.00	0.24	0.00	926.00
After	-75.00	0.00	0.00	-0.25	0.00	75.00

Table 5.1: Distribution of scopes 1 & 2 performance before and after outliers deletion

Table 5.1 shows how the distribution of scopes 1 & 2 performance has changed. A different approach to interpreting an emission trend is to calculate how many years the company submission has been compliant with scientific target-setting. For this purpose, the two-degree investment initiative offers seven different ways to calculate the compliance of data with different requirements on scope and data availability. Based on limited data availability in the CDP setting, this quantitative analysis uses data drawn only from scopes 1 & 2. Each of the seven approaches focuses on either the absolute or relative trend in emission reduction. This research measures the absolute emission trend in terms of the number of years compliant to the three-percent solution developed by the CDP and the WWF; to measure the relative emission trend, it considers the number of years compliant to the Climate Stabilization Intensity measure (CSI) (Tuppen, 2008; UNE, 2015; Hoehne et al., 2015a).

The three-percent target was developed based on US corporate emissions; the WWF and CDP calculated absolute emission-reduction targets, which depend on the years in which emission reductions would begin. E.g. if companies started reducing emissions in 2010, the required annual reduction is

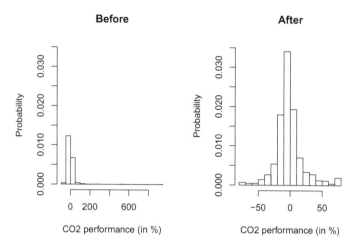

Figure 5.3: Distribution of Scopes 1 & 2 performance before and after outliers deletion, excluding missing and 0

3.2% for the years 2010-2020 and 4.3% for the years after. If companies were to begin reducing emissions in 2020, the annual efficiency increase needs to be 9.7%, and companies starting in 2030 must be immediately carbon neutral (Tcholak-Antitch et al., 2013, p. 29). No matter what emission reduction principle is applied, it is clear that companies that start reducing their carbon footprint earlier must have lower annual emission targets than companies that start later on. As the sample data include only data from 2010 to 2015 for scopes 1&2, the annual required increases are calculated depending based on the starting years between 2010 and 2015. As these are required annual decreases, the total reduction factor is the factor used to multiply base-year emissions. Therefore, the starting year is dynamically applied, while the end year is always 2015.

Compared to this absolute emission-reduction approach, *the Climate Stabilization Intensity (CSI) approach associates an organization's total carbon emissions with the contribution its profits and employment costs make to the world economy. Targets for reducing the company's carbon intensity (CO_2e per unit of contribution to GDP) are then set in line with world targets to reduce CO_2e emissions per unit of GDP* (Tuppen, 2008, p.1).

Starting in 2008, the absolute CO_2 data needs to be reduced by 2.1% per year; starting in 2015 the annual absolute reduction would be 2.8%. If no change occurs, it is estimated that global emissions will increase by 1.2%

each year. Table 5.2 provides an overview of the total reduction in emissions

	Start year	Annual decrease	Total reduction factor
1	2010	2.30	0.87
2	2011	2.40	0.89
3	2012	2.50	0.90
4	2013	2.60	0.92
5	2014	2.70	0.95
6	2015	2.80	0.97

Table 5.2: Required relative CO_2 emission reduction by starting year

per year (column annual decrease) and the reduction factor from the base year through 2015.

However, Table 5.2 cannot be directly linked to company emissions, as it does not account for growth. This is especially important for fast-growing industries or economies. Therefore, the CSI introduces intensity KPIs compared to the GDP growth of the industry. Taking annual worldwide GDP growth from 1990-2007 into account, and factoring in inflation, the model estimates that the intensity KPI of each company has to be reduced by 9.7% per year. The intensity KPI is defined as *Value Added*, which can be calculated as revenue minus costs of purchased goods and services. The reflection of the outsourced goods and services helps companies focus only on their own emissions, and rejects the possibility of benefiting directly from outsourcing, even though outsourcing of carbon-intense steps in the value-creation process would still improve the KPI (Tuppen, 2008, p. 6-10). Similar approaches are the CSO carbon metric and GEVA methods to calculate science-based targets (Science based targets, 2016b). The CSO carbon metric is especially strong for longitudinal, long-term studies with a scope of more than ten years, as it continuously recalculates companies' required emission reductions based on changing input parameters. Additional performance variables are generated straight from the CDP questionaire. The self reported percentage change of emissions (multiplied by -1) represents the emission trend reported from the company. Additionally, the description of emission reduction initiatives and downstream emission reduction initatives are used as seperated measures. The respective data is binary in terms of has initatives in places yes/no. The existance if the company has also emission reduction targets in place is operationalized by counting the letters in the description of what kind of targets are in place. On a log scale the respective values range from zero to eight.

III. Environmental disclosure

There are also various ways to operationalize the quality and amount of environmental-disclosure information. One common option for operationalizing disclosure quality is to use the disclosure score of the CDP (Guenther et al., 2015; The, 2015b; CDP, 2015i; Vlugt et al., 2015). This score is designed to provide stakeholder audiences with an easy way to judge how good a company's disclosure quality is. However, as the CDP was quite successful in motivating increased numbers of companies and asking for more detailed information over time, there was a need to adopt the scoring methodology over time to enable stakeholders better to differentiate among the leading companies. From 2016 on, the disclosure score is also not available any longer, as the disclosure elements are now covered by the performance rating. These changes could have unintended effects on an unbalanced panel study. An alternative way is to operationalize the reporting principles of the reporting standard and review how well companies are able to meet those requirements. As the principles are supposed to be applied with equal weighting, the results of each principle are multiplied with all other other principles to reflect the interdependencies among these principles. To operationalize the reporting principles, the CDP questions that focus on certain areas of disclosure quality are grouped to the respective principles. The five GHGP reporting principles (WBC, 2004, pp. 6-9) can be measured with the following elements.

RELEVANCE can be operationalized by the binary element alist. *Alist* represents whether all material scope 3 categories are reported. COMPLETENESS can be measured by the check of whether all three scopes have a value greater than 0. Reported values are represented with a 1 value per scope and a 0 for each scope not reported. Therefore, the variable can take values between 0 and 3. CONSISTENCY over time can be assessed by the number of years the company disclosed information to the CDP. This enables stakeholders to review the answers over time. If a figure has increased over time, the variable is divided by the maximum number of years reported in this year. TRANSPARENCY is reflected by the disclosure of the reporting and disclosure standard used. There is also a measure of the extent to which uncertainties in emission levels are made transparent. The standard elements are calculated by adding results of binomial variables together. This starts with any assurance standard variable, which reviews whether a company has answered the question of assurance standard for any scopes and counts the respective characters after excluding phrases for others or different spellings of inapplicable information. Application of ISO 14064, ISAE 3410 and AA1000AS is assigned a full point, while the respective less-specific standards ISO 140001, ISAE 3000, GHGP, GRI standard, and IPCC account for half a point. Compliance with a legal emission-trade or emission-tax obligation also counts for just for half a point. Transparency about remaining uncertainty is also calculated by adding binary variables for the transparency to data gaps, missing information, measuring issues, assumptions, sampling, emission factors and other data-management issues. Additionally, the numeric values of the uncertainty range and the amount of uncertainty description are scaled to a similar level and subsequently added. Afterwards, the columns for assurance standard and uncertainty are added. The ACCURACY principle is represented by the sum of the audit scores for scopes 1 & 2 and the audit score for all scopes. These scores are calculated by multiplying the assurance level per scope with the emissions per scope and dividing them by total emissions. As the results range between 0 and 200, both scores are applied on a log scale.

Additionally to these five GHGP reporting principles, the GRI has defined the three reporting principles of TIMELINESS, RELIABILITY and STAKEHOLDER COMMUNICATION. While reliability is covered by the audit score and timeliness is already partially operationalized in the consistency principle and requested in several reporting standards as well, which are included in the transparency principle, the STAKEHOLDER COMMUNICATION principle should be operationalized. Hence, access operationalizes stakeholders communication. This defines which user group has access to the data set. The respective calculations are performed via the mutate function in the dplyr package in R (Wickham and Francois, 2016; Peng and Matsui, 2016).

As Figure 3.5 in the model building section showed, different theories around voluntary carbon disclosure lead to different disclosure strategies. The main difference is between companies that want to let their stakeholders know that they are taking a proactive approach to handling climate change and providing high-quality reporting to send a positive carbon emission signal to the market, and companies that have no intrinsic motivation to disclose voluntary information and react to pressure to release the minimum amount of information to retain their license to operate from its stakeholders.

In between these extreme disclosure strategies, companies are also affected by isomorphic forces that shape homogeneous reporting practice. As the hypothesis aims mainly to differentiate between companies that report very well and all other companies, the disclosure score representing the different reporting principles can be applied on a log scale. Looking at the development of the application of the disclosure practice over time, the box-plot in Figure 5.4 shows two effects.

The first effect is that the leading group of the disclosure practice increases its reporting practice over time. This can be seen in the fact that the outliers on top are growing over time and achieving higher scores.

The second effect is that with reporting year 2015, the total box-plot moves up. This could be explained by the level of high public attention in 2016 to the ratification of the Paris agreement, as the CDP deadline for 2015 came in the middle of 2016. Hence, companies may have faced an increased interest on the part of their stakeholders, resulting either in a higher minimum requirement on disclosure or in stronger institutional forces. Table 5.3 shows

	2011	2012	2013	2014	2015
1	1467	1442	1472	1500	537
2	7	93	139	161	392
3	3	25	45	76	433
4	0	0	0	0	328
5	0	0	0	0	63
6	0	0	0	3	61

Table 5.3: Reporting principle score per year (log scale).

the respective plain numbers. Analyzing where the effects from the huge

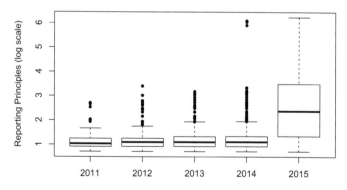

Figure 5.4: Change of conformity to reporting principles over time

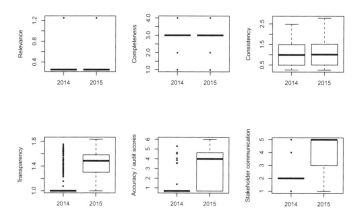

Figure 5.5: Breakdown of conformity to reporting principles over time

shift from 2014 and 2015 in the reporting principles are coming from, Figure 5.5 shows the operationalized principles of transparency, accuracy and stakeholder communication causing this effect, while relevance, completeness and consistency remain nearly constant. The transparency score is implemented from the results from the CDP questionnaire assurance standard as shown in CC8.6, along with information about uncertainty as defined in CC8.5 (CDP, 2014b, 2015f). Accuracy is the sum of the audit score for scopes 1 & 2(CC8.5 & CC8.6) and the audit score for all emissions (CC14.2) on a log scale, while stakeholder communication is the sum of access to the data and the identified risk categories (CC5 & CC6) (CDP, 2014b, 2015f). As all of these elements are available in both reporting years, the technical team of the CDP was interviewed to explore potential reasons for the changes, and to ensure the validity of the dependent variable over time. Table 5.4 shows the correlation between the dependent variable representing the disclosure practice and the independent variables representing CO_2 performance. As this simple correlation already shows, most of the performance variables correlate positively with the output variable. The negative correlation of the company's self-calculated performance stands in conflict with the hypothesis.

IV.   Mediator and moderator variables

Moderator effects accounting for SIZE, typically measured in terms of revenue, and industry SECTOR have a significant effect (Patten, 2002). The aim is thus

	Reporting Principles
Detail Targets	-0.12
Target category	0.48
Reduction Initiatives	0.50
Downstream Reduction	0.38
Absolute trend own emissions	-0.06
Compliant - 3 Degree target	0.27
Compliant - Climate Stabilisation Intensity	0.26

Table 5.4: Correlation between disclosure practise and carbon performance

to test the hypotheses including the moderator effect for size and industry. As the revenue size of the sample is biased by large, multinational companies, the revenue variable is used on a log scale. Incorporating the moderator

	significantCDP	significantGRI	significantGICS
Min. :	0.0000	0.0	-1.00
1st Qu.:	0.0000	0.0	0.00
Median :	0.0000	0.0	0.00
Mean :	0.0066	0.1	-0.07
3rd Qu.:	0.0000	0.0	0.00
Max. :	1.0000	1.0	0.00

Table 5.5: Summary of the significant effects coming from industry classifications.

variable of industry sector is more challenging. The reason for this is that industry clusters can be aggregated in several ways. Multiple suppliers for industry classifications, such as BIGS, GIGS, etc., exist. In addition, these classifications are done on multiple levels and can be aggregated respectively and are structured as factor variables. In a linear regression, the classical approach would be to transfer each industry sector to a binary variable and test for each industry separately to ascertain which of these industries has a significant positive or negative effect on the outcome variable. To minimize the analysis down to one line per classification type, the dependent variable is tested against all different industries on the lowest available level. The results are clustered into significant positive (+1), significant negative (-1) and non-significant (0) sectors. This classification is subsequently merged with the data set. As Table 5.5 shows, this is performed for the CDP activity clusters, GICS cluster and GRI sector definition. Each industry cluster has either positive or negative effects, never both. Also the number of companies with positive or negative effects is quite small as the absolute mean value indicates. Besides these moderator effects for size and revenue, there are further moderator effects that stem from the impact of the stakeholders. Here, the literature is currently not clear as to which moderator effects shall be taken into account and which not. As the theoretical concept of Chapter 3 implies, disclosure activity is a function of the expected green rents and stakeholder salience. While green rents are likely to be industry-specific and are hence covered under the industry classification.

Based on the signaling setting that companies want to send positive signals to satisfy various stakeholder requests, as shown in Figures 3.8 and 5.2, the respective stakeholder variables should function as mediator variable. To represent the importance of the company to environmental stakeholder, the GRI index is used. The GRI index represents all relevant indices in which a company is listed. If a company is listed in more than one index, both indices are named, separated by comma. The core assumption is that a company listed in multiple indices also faces more investor pressure to disclose information. This operationalization accounts for the fact that CDP reporting already has a strong investor bias, as all companies submitting data do so at the invitation of their investors. Another relevant stakeholder is the government. To assess the salience of the government in the field of GHG emissions, three parts are important. The first of these concerns how strictly GHG emissions are regulated. Here, the GHG politics score is used in research (Guenther et al., 2015). The GHG politics score is a result of interviews of NGO members in several countries. These NGOs assess local GHG politics and its effectiveness based on a structured questionnaire. The answers are calculated to a so called *GHG score*. The score is published by the NGO Germanwatch (Burck and Marten, 2016). As the total number of persons interviewed increases over time, and due to job fluctuation, the trend within a country can change over time. After a discussion with the Germanwatch experts, it was decided that the latest score would be used to assess all years. The next element is the effectiveness of law enforcement within a country. This law enforcement is measured by the World Bank on an annual basis (Indicators, 2016; Kaufmann et al., 1999). As carbon regulation has to be seen in the context of local regulation, it is expected that common-law countries will have less detailed regulation, in turn affecting the perception of regulation and also the effectiveness of the government. Therefore, the binary variable for the common-law country has to be added. As regulators and investors often focus on the highest-emitting polluters, the variable of CO_2 intensity clusters the companies into CO_2 intensive and non-intensive sectors.

The next stakeholder group is the public, which comprises employees, customers, NGOs and other relevant stakeholder groups. These groups are represented by the ability of the media to influence the legislative process. The respective operationalization factor is also taken from the World Bank and is named there as *Voice and Accountability* (Indicators, 2016). As there are also other local stakeholder effects that might not be covered by the respective operationalization figures, a variable representing the headquarters country of the reporting company is also added. This country list was also clustered into significant positive/negative and non-significant countries.

C.   Descriptive statistics

Table 5.6 shows the distribution of each variable used to test the first and third hypothesis. The first variable represents the dependent disclosure variable for the first hypothesis, the second variable the dependent variable of the third hypothesis. The following seven variables are the operationalized values of the carbon performance. The remaining eleven variables represent the mediator and moderator effects. The dataset contains 1301 observations. All short and long descriptions for the variables can be found in the attachment in Table 6.3. Short variable names beginning with H represent the dependent variable in the respective hypothesis. The P stands for the variables connected to carbon performance, while the M represents mediator and moderator variables. The preparation of the data set and the statistical tests were performed with the R version 3.3.2 on a 64-bit Linux platform using RStudio version 1.0.136 (Aphalo, 2016; Caffo, 2015; Dolic, 2004; Crawley, 2012; Ligges, 2006; R Development Core Team, 2008). To keep the research reproducible, the chapter and the code are written in one R-Markdown file and then converted into Latex (RStudio Inc., 2016). The tables are formated with the R packages stargazer (Hlavac et al., 2015; Hlavac, 2015), pander (Daróczi and Tsegelskyi, 2016), and xtable function from the xtable package (Dahl, 2016).

Table 5.7 provides a breakdown by CDP industry cluster for the year 2015. Companies from the sectors Services, Maunfacture, Infrastructure and Materials dominate the total sample by representing 74% of the company submissions. While Services are considered as low CO_2 intense sectors, Man-

Statistic	N	Mean	St. Dev.	Min	Max
Reporting principles (log scale)	1,301	2.505	1.283	0.715	6.213
S12auditscore	1,301	62.074	67.173	0.000	200.000
Compliant - 3-Degree target	1,301	2.724	2.386	0	5
Compliant - Climate Stabilisation Intensity	1,301	2.918	2.391	0	5
Self-reported trend	1,301	−0.508	19.207	−75.000	75.000
Emission-reduction initiatives	1,301	0.766	0.424	0	1
Downstream reduction initiatives	1,301	0.576	0.494	0	1
Target category	1,301	0.988	0.908	0	3
Detail Abs Target	1,301	0.942	2.294	0	8
GHG politics	1,301	53.548	10.653	37.230	71.190
Legal system	1,301	0.367	0.482	0	1
Public voice	1,301	0.932	0.716	−1.657	1.710
Government effectiveness	1,301	1.314	0.652	−0.277	2.194
Intensity factor	1,301	0.317	0.465	0	1
Revenue (log scale)	1,301	9.899	2.724	2.864	19.117
GRI index listing	1,301	41.872	28.339	3	237
Significant CDP	1,301	0.007	0.083	0	1
Significant GRI	1,301	0.107	0.309	0	1
Significant GICS	1,301	−0.071	0.258	−1	0
Significant country	1,301	−0.778	0.416	−1	0

Table 5.6: Summary of the input variables used to test Hypothesis 1 and 3 with 2015 data

ufacturing, Infrastructure and Materials are associated as emission intense sectors. Comparing the median values by sector with the standard deviation shows that the majority of companies submit data to the CDP which is not consistent with the general reporting principles of the GHGP. This distribution is consistent with the information given during the qualitative study that a majority of companies claims to be green but does not report the respective information in a reliable manner. Geographically the headquarters of the reporting companies are distributed across all continents. Most often the reporting companies are based in USA, Japan, United Kingdom, Canada and China. The detailed breakdown by countries can be found in the attachment on Table 6.1. Table 6.4 shows the CORRELATION between the variables. Both disclosure variables correlate between 0.38 and 0.52 with the performance variables for emission reduction initiatives (P4), the downstream reduction initiatives (P5), and the reduction targets (P6). The correlation coefficient between both dependent variables is with 0.7 highly supporting hypothesis 3. The correlation between the disclosure variables and the moderator and mediator effects is low, supporting the theoretical construct that high quality voluntary disclosure can be explained by agency theory in contrast to external pressure. Looking into the performance variables the correlation (0.59) between the 3-degree target (P1) and the CSI compliancy target (P2) can be easily explaied as both scores are calculated based on the own emission trend. The correlation between the emission reduction initiatives for own (P4) and downstream emissions (P5) and the companies targets on emission reduction (P6) is also logical. Unexpected is that the self reported emission trend (P3) is not correlated to two-degree targets (P1 & P2) even though these are calculated based on emission trends reported to the CDP. This mismatch is consistent with information from the interviews that companies do not report consistently across channels. Own emission reporting showing a positive emission trend could be also potentially seen as green washing element, as it does not correlate with any emission reduction activities (P4, P5, P6). Also providing very detailed description of the targets does not correlate with the disclosure variable nor the audit variable or any othe performance variable. This seems to be also an indicator for green washing. As Figure 5.6 shows, depending on the number of compliant years, the disclosure quality measured by the reporting principles increases. The quality increase between the number of compliant years is a poor measure to separate between two-degree-compliant companies and those that are not compliant, as the huge overlaps in the field of accounting quality shows. A comparison between 2014 and 2015 indicates that the the overall disclosure practice increases and that

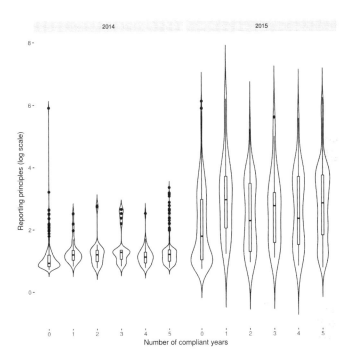

Figure 5.6: Reporting practice by number of compliant years.

CDPindustry	N	in %	Mean	SD	Min	Median	Max
1 Services	391	30.00	30	63	0.04	10	434
2 Manufacturing	369	28.00	30	72	0.13	6	448
3 Infrastructure	104	8.00	33	65	0.13	16	386
4 Materials	100	8.00	31	70	0.26	12	435
5 Biotech/Health Care/ Pharma	94	7.00	17	23	0.09	6	151
6 Fossil Fuels	88	7.00	24	42	0.2	7	253
7 Transport	66	5.00	31	66	0.26	12	355
8 Power Generation	44	3.00	30	72	0.13	12	482
9 Mineral Extraction	36	3.00	58	104	0.26	14	497
10 Food/Beverage/ Agriculture	4	0.00	2	3	0.53	1	7
11 Hospitality	2	0.00	25	34	0.53	25	49
12 Retail	2	0.00	13	11	5.78	13	21
13 Apparel	1	0.00	19		18.8	19	19

Table 5.7: Reporting principles by CDP industry mapping

a significant portion of the non-compliant companies provides high quality reporting. In both years, the mean of the reporting principles is lowest in the cluster without any compliance with the science-based trend of emissions, which could be an indication for the signaling function of a high-quality carbon disclosure.

D. Environmental performance and comparable disclosure

The first hypothesis is defined as:

Companies with positive environmental performance disclose better comparable information than companies with less good performance.

The hypothesis test for Hypothesis 1 can be performed with a simple general model on a single-year level or for all of the panel data with a respective panel test. The first test is performed with the single-year test using data for 2015. All variables are designed such that a positive sign supports the hypothesis, while a negative sign shows an effect contradictory to the hypothesis. Table 5.8 shows that not all variables have significant effects on

| | *Dependent variable:* | | |
	Reduced	All	interactions
P1	0.038*** (0.014)	0.038*** (0.014)	
P2	0.065*** (0.014)	0.066*** (0.014)	
P3		−0.002 (0.001)	
P4	0.858*** (0.093)	0.843*** (0.098)	0.997*** (0.085)
P5	0.300*** (0.070)	0.264*** (0.073)	
P6	0.327*** (0.038)	0.362*** (0.045)	−0.469** (0.207)
P7		0.028* (0.015)	
M1		0.002 (0.004)	
M2		−0.040 (0.074)	−1.136*** (0.289)
M3	0.276*** (0.062)	0.253*** (0.066)	
M4	−0.196*** (0.063)	−0.168** (0.067)	
M5		0.008 (0.066)	
M6	0.029** (0.012)	0.035** (0.015)	
M7	0.004*** (0.001)	0.004*** (0.001)	
M8	0.858*** (0.331)	0.890*** (0.331)	
M9	0.331*** (0.091)	0.326*** (0.092)	
M10		0.122 (0.114)	
M11	0.426*** (0.078)	0.401*** (0.082)	
P1:P2			0.014*** (0.002)
M2:P3			−0.004* (0.002)
M3:P5			0.133*** (0.042)
P5:M9			0.669*** (0.145)
P6:M1			0.008*** (0.003)
P6:M6			0.041*** (0.010)
M2:M6			0.116*** (0.033)
M9:M5			0.838*** (0.200)
Constant	0.904*** (0.166)	0.662** (0.316)	1.181*** (0.063)
Note:			*p<0.1; **p<0.05; ***p<0.01

Table 5.8: Hypothesis 1 tested with 2015 data 1/2

Observations	1,301	1,301	1,301
R^2	0.419	0.422	0.399
Adjusted R^2	0.413	0.414	0.394
Residual Std. Error	0.983 (df = 1288)	0.983 (df = 1282)	0.999 (df = 1289)
F Statistic	77.375*** (df = 12; 1288)	51.946*** (df = 18; 1282)	77.869*** (df = 11; 1289)
Note:			*p<0.1; **p<0.05; ***p<0.01

Table 5.9: Hypothesis 1 tested with 2015 data 2/2

the dependent variable representing disclosure. Therefore, the model can be simplified by eliminating variables P3 (Self reported trend), P7 (Detail Abs Target), M1 (GHG politics), M2 (legalsystem), M5 (Intensity factor), M10 (significantGICS) which have not significant effects on the hypothesis (differences between column "Reduced" and "All" in Table 5.8). Column "interaction" includes also all signinficant effects from interaction variables. In all three columns it can be seen that the self reported trend (P3) and the self described details on the targets (P7) have no effect on the disclosure quality, while emission reduction over time in the CDP (P1, P2) and reduction activities (P4, P5, P6) do have a significant positive effect on the disclosure quality. Partially this effect is supported by the GHG politics in the country (M1), Public voice (M3) or the company size (M6).

	Df	Sum Sq	Mean Sq	F value	Pr(>F)
P1	1	149.48	149.48	154.84	0.0000
P2	1	36.09	36.09	37.38	0.0000
P3	1	2.44	2.44	2.53	0.1100
P4	1	506.37	506.37	524.53	0.0000
P5	1	39.37	39.37	40.78	0.0000
P6	1	80.03	80.03	82.90	0.0000
P7	1	1.88	1.88	1.95	0.1600
M1	1	0.01	0.01	0.01	0.9200
M2	1	7.52	7.52	7.79	0.0100
M3	1	0.08	0.08	0.09	0.7700
M4	1	5.12	5.12	5.31	0.0200
M5	1	0.58	0.58	0.60	0.4400
M6	1	15.15	15.15	15.70	0.0000
M7	1	9.52	9.52	9.86	0.0000
M8	1	8.33	8.33	8.63	0.0000
M9	1	16.41	16.41	16.99	0.0000
M10	1	1.41	1.41	1.46	0.2300
M11	1	22.86	22.86	23.68	0.0000
Residuals	1282	1237.61	0.97		

Table 5.10: Anova Hypothesis 1 tested against reporting principles

The anova function supports this trend. To additionally cover potentially relevant interactions among the variables, the reduced model is compared with a model incorporating the interactions of all variables, reduced to only relevant variables and relevant interactions of all variables. Even though the new model is more complex by incorporating more variables and its combinations, the overall r-squared and the respective adjusted r-squared are decreasing. Hence, the simplified model can be used as core model to interpret Hypothesis 1. The interpretation of the mediating effects of the stakeholder variables will be discussed in C-Hypothesis 4 and 5.

As all performance variables have a significant positive effect on the disclosure variable with a significance level of $p < 0.01$, the Null hypothesis - no effect- can be rejected. Or, put the other way round, there is strong evidence that carbon performance and carbon disclosure are positively correlated in the sample for 2015. The next section checks whether this connection can also be seen in a five-year panel of data from 2011 through 2015.

The model selection for this PANEL MODEL follows the guideance of the Econometrics Academy (Katchova, 2013). Hence, the pooled, fixed effects and random-effects models are calculated. If more than one model is fitting, the last named model is applied. To identify the true model, the between estimator and the first differences estimator are calculated as well. The calculation is performed using the *plm* function in the *plm* package (Croissant et al., 2016). Table 5.11 shows the respective overview of which results should be

Estimator/true-model	Pooled-	Random effects-	Fixed effects-
Pooled OLS	Consistent	Consistent	Inconsistent
Between	Consistent	Consistent	Inconsistent
Within or fixed effects	Consistent	Consistent	Consistent
First differences	Consistent	Consistent	Consistent
Random effects	Consistent	Consistent	Inconsistent

Table 5.11: Models and estimators

consistent/inconsistent to each other in order to apply the respective model.

These five models/estimators are compared for consistent and inconsistent effects to determine which model is the true model. The first comparison is done within the pooling model to check if random effects occur over time. This is done with the Breusch-Pagan Lagrange Multiplier test for random effects over time. As shown in Table 5.12, one model is inconsistent as there are significant effects between the models. Therefore, it is recommended to use the random- or fixed-effects model, instead of the the pooling model

statistic	p.value	alternative
17.73	0.00	significant effects

Table 5.12: Lagrange Multiplier Tests for Panel Models with OLS

(Katchova, 2013). The decision between the fixed-effects model and the random model can be done by interpreting the Hausman test (Katchova, 2013) (Torres-reyna, 2010, p. 16). As shown in Table 5.13, the p-value is < 0.05, indicating that the fixed-effects model should be used. Applying

statistic	p.value	parameter	alternative
111.95	0.00	9	one model is inconsistent

Table 5.13: Hausman Test for Panel Models between random model and fixed-effects model

the fixed-effects model to the data set, most variables representing carbon performance have a positive significant effect on carbon disclosure, as shown in Table 5.14. The reduced number of variables owes to the fact that only time-dependent variables are used in the fixed-effects model calculation (Katchova, 2013). As in the 2015 sample, the mediating effect of revenue shows a significant negative effect, while all other stakeholder variables are not attached to any significant effect.

Splitting the data set into a group of the highest transparency within each year, represented in column *Highend* which is the 10% quantile of the disclosure variable reporting principles, it can be seen that the R squared for the best disclosing companies increases and at the same time another element representing carbon performance turns significant, and the unexpected revenue effect loses its significance. Both effects strongly support the agency hypothesis that carbon disclosure is done especially well to enable stakeholders to identify the company as green.

The same test can be performed for all company submissions with a carbon disclosure worse than the median. The results are shown in column *Tail* of the same table. Under this approach, the r-squared is smaller than in both other models, which already indicated that the correlation between performance variables and disclosure is not as highly developed. Consistent with the correlation and the 2015 test is that self-reported performance has no effect on the disclosure quality, the emission reduction activities and the target category has implications while the details on the target have no impact (except comparing the high end with all other companies) on the disclosure quality.

With the connection between carbon performance and carbon Disclosure demonstrated, the next hypothesis considers to what degree high-quality disclosure is also connected to meeting the two-degree target.

E. Positive CO_2 disclosure and two-degree target

The majority of companies which send a positive CO_2 disclosure signal meets the scientific two-degree target.

Since this hypothesis is formulated based on an absolute measure, the test of the hypothesis can be done based on descriptive statistics. As the two-degree compliance was calculated in the data set based on the year 2015, the analysis will be done with the 2015 submission only.

First, the data are split into two sets: the leadership group and all other submissions. As Figure 5.7 indicates, the majority of companies with high-quality reporting are two-degree compliant if only scopes 1 & 2 emissions are considered. The table shows 2014 and 2015 single-year data. Column Abs_S12, represents the years in which the absolute scopes 1 & 2 trend is compliant with the absolute two-degree emission reduction trending as defined by CDP & WWF (Tcholak-Antitch et al., 2013). The relative CSI trend is represented in column Rel_S12 (Tuppen, 2008). In addition, the leadership group is more often two-degree compliant. However, reviewing this hypothesis for all scopes, almost no companies can be two-degree compliant on absolute measures, as shown in column Abs_all. Interestingly, in 2014, all compliant companies were part of the disclosure leadership group, while in 2015 disclosure practices improved, and the the leadership group has a compliance distribution similar to that of the overall population.

The difference between the columns Abs_All and Abs_S12 in Figure 5.7 suggests that almost no company is two-degree compliant over the full life cycle. Due to the small number of companies that are visibly two-degree compliant over the full value chain, a quantitative assessment of the indicators is not possible with the data currently available. This statement can be seen even more clearly in Figure 5.8, as the number of companies not coded in red cannot easily be spotted.

	Dependent variable: log_reporting_principles		
	Full data	Highend	Tail
Reported own	−0.001	−0.003	−0.0001
performance	(0.001)	(0.002)	(0.001)
Downstream Reduction	0.369***	0.468***	0.143***
	(0.042)	(0.099)	(0.034)
Emission-Reduction	0.948***	1.899***	0.784***
Initiatives	(0.051)	(0.116)	(0.041)
Target Category	0.323***	0.108**	0.278***
	(0.027)	(0.054)	(0.024)
Detail Abs. Target	0.007	0.045**	0.008
	(0.008)	(0.021)	(0.007)
GHG Politics	0.054	−0.062	0.040
	(0.085)	(0.172)	(0.059)
Public Voice	0.154	0.241	−0.173
	(0.188)	(0.361)	(0.148)
Government	0.092	−0.125	0.215***
Effectiveness	(0.097)	(0.207)	(0.073)
Revenue (log scale)	−0.121**	−0.010	−0.032
	(0.049)	(0.099)	(0.037)
Constant	0.062***	0.177***	0.044***
	(0.010)	(0.029)	(0.009)
Observations	4,579	684	3,118
R^2	0.572	0.798	0.595
Adjusted R^2	0.571	0.795	0.594
F Statistic	677.252***	295.584***	507.853***
	(df = 9; 4569)	(df = 9; 674)	(df = 9; 3108)
Note:			*p<0.1; **p<0.05; ***p<0.01

Table 5.14: Hypothesis 1 tested with Panel data for years 2011 - 2015

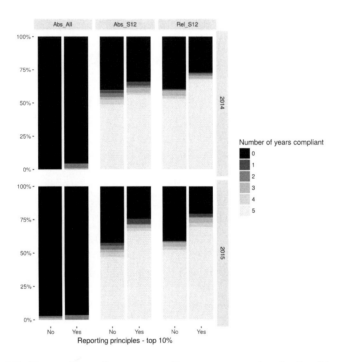

Figure 5.7: Years of two-degree compliance compared to leadership group of disclosure 2014 & 2015.

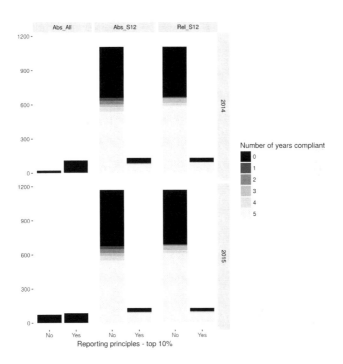

Figure 5.8: Years of two-degree compliance compared to leadership group of disclosure 2014 & 2015, absolute figures.

F.   Environmental performance and voluntary audit services

The measures representing carbon disclosure and carbon performance were often perceived as poor in the literature review. As Hypothesis 1 works with a complex but effective measure to operationalize carbon disclosure, Hypothesis 3 now tests whether a shortcut to the extent of using a voluntary audit could also serve as an operationalization factor. In a first step this is being done with 2015 data. Having proofed in Hypothesis 1 that disclosure practice and performance correlate, we assume that if we find a similar effect with the dependent variable of the audit score, the audit score can be used to operationalize disclosure. With this in mind, the relevant hypothesis is as follows:

> *Companies with positive environmental performance use voluntary audit services more often than companies with a poor performance do.*

This hypothesis is a triangulation of the first hypothesis, as voluntary audit service is a generally accepted principle to support arguments of the agency theory. To test the hypothesis, the reporting principle variable is replaced with the audit score variable (both on a log scale). As with the first hypothesis, the hypothesis is first tested with 2015 data only. Also, the simple linear model and same variables are used.

As the interaction results did not provide a better model, the general models for Hypotheses 1 and 3 can be directly compared with each other. The results for this hypothesis test can be seen in Table 5.15. Looking at the results, most variables have a similar outcome and significance as in Hypothesis 1. Consistently the self reported emission trend (P3) and the target details (P7) have no impact on the disclosure strategy. While the absolute measure for the two-degree target used to have a significant impact, it looses its significant effect with the audit score as dependent variable. GHG Politics (M1) and Legal system (M2) gain significant effects. The overall r-squared even increases, pointing to the assumption that the audit score for scopes 1 & 2 can be used to operationalize carbon performance. Table 5.16 shows the anova table comparing the outcomes of the model used for Hypothesis 1 and Hypothesis 3.

As for Hypothesis 1, the approach from Katchova (2013) is followed to choose the right PANEL MODEL. With these five models, in a first step the pooling model is tested with the Breusch-Pagan Lagrange Multiplier test for random effects over time. As shown in Table 5.17, one model is inconsistent.

	Dependent variable:	
	Hypothesis 3	Hypothesis 1
	Audit score	Reporting principles
P1	0.033 (0.028)	0.038*** (0.014)
P2	0.060** (0.028)	0.066*** (0.014)
P3	−0.004 (0.003)	−0.002 (0.001)
P4	1.531*** (0.190)	0.843*** (0.098)
P5	0.319** (0.140)	0.264*** (0.073)
P6	0.774*** (0.088)	0.362*** (0.045)
P7	0.026 (0.028)	0.028* (0.015)
M1	0.017** (0.007)	0.002 (0.004)
M2	−0.280** (0.142)	−0.040 (0.074)
M3	0.318** (0.127)	0.253*** (0.066)
M4	−0.309** (0.129)	−0.168** (0.067)
M5	0.087 (0.128)	0.008 (0.066)
M6	0.071** (0.029)	0.035** (0.015)
M7	0.009*** (0.002)	0.004*** (0.001)
M8	1.500** (0.639)	0.890*** (0.331)
M9	0.470*** (0.178)	0.326*** (0.092)
M10	0.182 (0.219)	0.122 (0.114)
M11	0.722*** (0.159)	0.401*** (0.082)
Constant	0.075 (0.609)	0.662** (0.316)
Observations	1,301	1,301
R^2	0.402	0.422
Adjusted R^2	0.394	0.414
Residual Std. Error (df = 1282)	1.895	0.983
F Statistic (df = 18; 1282)	47.931***	51.946***
Note:	$^*p<0.1$; $^{**}p<0.05$; $^{***}p<0.01$	

Table 5.15: Hypothesis 3 compared to Hypothesis 1 with 2015 data

	Res.Df	RSS	Df	Sum of Sq	F	Pr(>F)
1	1284.00	4762.22				
2	1282.00	1237.61	2.00	3524.61	1825.51	0.00

Table 5.16: Anova comparing results from Hypothesis 3 with results from Hypothesis 1

Use of the random model instead of the pooling model is thus recommended (Katchova, 2013).

statistic	p.value	alternative
8.79	0.00	significant effects

Table 5.17: Lagrange Multiplier Tests for Panel Models with OLS for Hypothesis 3

statistic	p.value	alternative	df1	df2
0.24	1.00	significant effects	1309	4569

Table 5.18: F-Test for individual and/or time effects between fixed-effects model and pooling model for Hypothesis 3

Since it has been shown that there are time effects, it has to be decided, if a fixed-effects model or a random model is preferable. To test this, the Hausman test (Katchova, 2013) (Torres-reyna, 2010, p. 16) is recommended. As shown in Table 5.13, the p-value is > 0.05, indicating use of the random model. Applying the fixed-effects model to the data set, most variables representing carbon performance have a positive significant effect on carbon disclosure, as shown in Table 5.14. The mediating effect of revenue is unexpectedly negative. All other stakeholder variables are not attached to any significant effect. Splitting the data set into a group of the highest transparency within each year, represented in column *Highend*, which is the 10% quantile of the disclosure variable reporting principles, it can be seen that the R squared for the best disclosing companies increases, while at the same time another element representing carbon performance turns significant and the unexpected revenue effect loses its significance. Both effects strongly support the

statistic	p.value	parameter	alternative
11.59	0.24	9	one model is inconsistent

Table 5.19: Hausman Test for Panel Models between random model and fixed-effects model for Hypothesis 3

agency hypothesis that carbon disclosure is done especially well to enable stakeholders to identify the company as green.

The same test can be done for all company submissions with a carbon disclosure worse than the median. The results are shown in column *Tail* of the same table. In this approach the r-squared is smaller than in both other models, indicating already that the correlation between performance variables and the disclosure is not as pronounced. Moreover, the variable about details on target increases its significance, while size is not significant. Supporting the legitimacy theory, the test increases government effectiveness for the first time. With the check for the connection between carbon performance and carbon disclosure, the next hypothesis reviews the extent to which high-quality disclosure is also connected to meeting the two-degree target.

Transferring this test to a PANEL ANALYSIS is shown in attachment Table 6.6. For the full data set, the absolute scopes 1 & 2 compliance measurement, the downstream and general reduction initiatives, and the target category influence the third-party verification which is used to measure disclosure. Stakeholder legal system and government effectiveness have a negative effect on voluntary disclosure. Government effectiveness in particular is unexpected here, as it is associated with higher governmental pressure and therefore associated with increased voluntary disclosure within the literature. Comparing the full data model between Hypothesis 3 as shown in the second column and Hypothesis 1 as shown in the third column, we find consistency in the effects of the performance variables. The variables with a significant effect vary depending on which data set is used. Consistent with Hypothesis 1 and Hypothesis 3 in the single year is, that the performance variables P4, P5 and P6 (Highend sample slightly different) have a sginificant effect on discolure. The self reported emission trend and the self reported target description has no significant impact on the disclosure quality. The audit score responds more to stakeholder variables than an overall disclosure variable. This could explain the increased r-squared of the model for Hypothesis 3.

The unavailability of scope 3 emissions was identified as a limitation of the results for Hypothesis 1. Additionally, Figure 5.7 shows that the two-degree compliance statements are only valid for the own emissions with limitations on the overall assessment of the validity, as the respective measures needed to operationalize a consistency of the value add of each company are lacking. Having seen that an audit score is a similarly good measure to represent carbon disclosure associated with carbon performance, the next descriptive statistic tries to narrow down whether a scope 3 audit score could be a measure representing two-degree compliance over the full value chain. Hence, filtering the companies down to company submissions which provide a sufficient amount of information on scope 3 transparency, it can be seen that these companies do not comply with the two-degree target on all scopes, no matter if the scope 3 emissions are audited or not. This indicates that, for this sample, scope 3 assurance is currently no indicator for carbon performance that meets the two-degree target for the full value chain. The next two C-hypotheses review the extent to which the expected mediating effects from the literature demonstrated a significant impact on the companies' disclosure practices.

G. *Moderator and Mediator effects on disclosure*

I. Stakeholder salience

Companies voluntarily disclose information as a result of stakeholder pressure. The amount & quality of environmental disclosure is a function of stakeholder salience.

This hypothesis stemmed from conflicting empirical evidence from Guenther et al. (2015), who demonstrated that there are significant effects based on stakeholder salience. As this analysis uses CDP data, it is expected that similiar effects are applicable in this thesis, too. In contrast to this, Patten indicated that, besides size and sector, no further moderator variables proved significant over a range of empirical works (Patten, 2002, 2015).

The results from Hypotheses 1 and 3 filter down to the stakeholder variables as shown in Table 5.20 To understand the impact of the stakeholder variables, Table 5.20 provides an overview of the respective significance. One is connected to a positive significance, zero to no significance, minus to significance with negative effect, and NA means that the variable could not be used in the fixed-effects model, as the variable is not calculated/defined in a way that it changes its values over time. The respective details are shown in the attachment on Tables 6.8 and 6.9. Looking into the different stakeholder

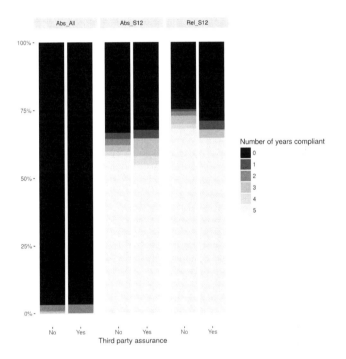

Figure 5.9: Years of two-degree compliance depending on third-party verification of scope 3 emissions.

Stakeholder variable	H1 2015	H3 2015	H1 Panel	H3 Panel
GHG politics	0	1	0	0
Legal system	0	-1		-1
Public voice	1	1	0	1
Government effectiveness	-1	-1	0	-1
Intensity factor	0	0	-1	0
Revenue (log scale)	1	1		0
GRI index listing	1	1		1
Significant CDP	1	1		1
Significant GRI	1	1		0
Significant GICS	0	0		0
Significant country	1	1		1

Table 5.20: Mediator and moderator effects of stakeholder variables

variables, it becomes clear that the GHG politics score has only a limited impact. This is potentially caused by the fact that the only globally available KPI, which measures GHG politics, is based on qualitative NGO interviews, which reflect the respondent's own local view of government regulative intensity. This could limit the measure effect of GHG politics. Future research could potentially explore whether KPIs measuring enforcement of global agreements on climate (OECD, 2015) might be better. Unfortunately, these measures are currently not available. Common-law countries seem to have a negative effect on carbon verification of companies but no measurable effect on all disclosure elements. A strong public voice, weak law enforcement, and financial listing also seem to positively influence the disclosure practice of companies. While the effect of a strong public voice and increased stakeholder interest suits the general reasons for voluntary disclosure, weak government effectiveness can be interpreted in two ways. One potential explanation could be that companies feel a greater need to provide voluntary information to stakeholders, as government is not requesting it. Alternatively, companies could be motivated to disclose more potentially incorrect information to stakeholders as they feel a weaker litigation risk. This rather negative explanation, however, conflicts with the setting of the dependent variable, which under both hypotheses also includes the amount of audited information. Interestingly, the relevance of government effectiveness turns out to have positive significance for disclosure in the panel data set, if only companies with a reporting factor worse than the median reporting factor are taken into account, as Table 5.14 shows. For the audit score panel analysis, the governmental effect loses at least its unexpected negative significance and turns out to be insignificant.

It is also interesting that the intensity factor seem to have no/negative effect on the disclosure practice of companies. This is also unexpected. Expected, however, is, that the country in which the headquarters are located has a significant effect on the disclosure practice of companies. This variable was included in the data set to ensure that country specifics not covered by the moderator and mediator variables used could be covered.

Recalling the interaction effects as shown in the third column of Table 5.8. All single-stakeholder effects lost their significance, with the exception of the legal system, which was not significant before and is now significantly negative; this is potentially explainable based on the positive significance with the interaction variables. This indicates that stakeholder effects might be further analyzed in interaction effects.

Increased GHG regulations have a significant effect on disclosure in combination with company target setting. Not unexpectingly legal system and/or intensity in interaction with government effectiveness leads to significant results. That a strong public voice in combination with climate change actions by companies is associated with voluntary disclosure is not be surprising. The same applies to the interaction between GHG politics and target setting and green initiatives and the respective association with disclosure.

Significant interaction between the variables of revenue and government effectiveness is not surprising, as the literature states that revenue and sector have significant impacts on disclosure practices.

II.   Company size

Big companies disclose more information than small companies.

This control-hypothesis was proved correctly in a literature review by Patten focusing on quantitative environmental disclosure studies (Patten, 2002, 2015). Looking at all disclosure variables, the relationship between size and disclosure practice could be confirmed. Looking only at the assurance practice of companies, there was no significant correspondence of size in the respective models. As the CDP sample includes mainly big multinational companies, the explanation factor of this confirmatory hypothesis is limited.

Interesting to acknowledge is that the interaction variable of size and target setting has shown significant effects on disclosure as well as the interaction variables of legal system and size. The other generally accepted moderator variable in disclosure research is the sector classification.

III.   Sector

Amount of disclosure is dependent on the sector of the company.

As shown in Table 5.20, CDP and GRI sector mapping had significant effects on disclosure practice, while the GICS classification did not have an effect. This difference could be caused by the different level of granularity, as the CDP activities were split into 142 levels, the GRI into 48 levels compared to the broader breakdown of eleven GICS clusters. Alternatively, it could be caused by scoping differences, as GRI and CDP try to cluster around similar environmental (CDP) or ESG (GRI) criteria; this was also pointed out in the interviews.

Chapter 6: Recapitulation and future prospects

A. Results and implications

I. Indicators for green-washing

Based on the theoretical model adopted from the agency theory as shown in Figure 3.8, a research gap was identified on how companies could send green signals based on their carbon inventory and how those signals could be interpreted by stakeholders to distinguish between truly green companies considered as cherries and green-washing companies considered as lemons. This differentiation is especially important as the literature review showed that green-washing strategies are chosen on a regular basis by companies and current measures to operationalize green performance in research do not permit identification of green-washing companies.

As the theoretical model builds upon the agency theory and the signaling argument, this research faces the challenge of proofing whether the side conditions of the signaling argumentation can be fulfilled. As the literature review showed that there are certain investors and other stakeholders willing to invest into truly green companies or support truly green companies in other respects, the remaining side conditions are that the company must convince stakeholders that the company is truly green. As most companies potentially have an incentive to convince their stakeholders that the reporting company is truly green, the truly green companies shall send a signal which cannot be copied by less-green companies or that can only be copied at significantly higher cost. Additionally, stakeholders have to be able to interpret the different signals to differentiate between green and non-green companies.

As the calculation and respective interpretation of CO_2 inventory information is quite complex, the ability to interpret the respective signals shall be analyzed taking into account the underlying limitations of the reported information. To understand the limitations of the data and the remaining ability to interpret the information, eleven subject-matter experts were interviewed. The aims of the interviews were to understand how inventory disclosure can be interpreted and what information companies shall provide to increase the comparability of the carbon inventory. The answers of the interviews can be also negatively interpreted, as companies which voluntarily decide not to disclose information that increases comparability do not aim to achieve comparability. This decision can be associated with a disclosure strategy not to enable stakeholders to distinguish between cherries and lemons, which is considered a green-washing (lemon) strategy.

According to the interviews, indicators that the reporting company strives for comparability include that companies report consistent figures over all distribution channels. This seems to be an obvious requirement; however, the feedback was that companies often use different CO_2 inventory numbers over various distribution channels. This limits the reliability of all numbers and adds complexity to any comparison. As gathering a carbon inventory is quite complex, it needs to be consistent with regard to boundaries, scope, methods and other reporting elements over time. Alongside consistent reporting, it is important to ensure the completeness of the report. Completeness is defined by the experts as a carbon inventory including all material emissions over all scopes. Like consistency, completeness is also a main principle in several reporting standards. It is thus surprising that the experts indicate that the majority of companies fail to provide information which follows this guidance, according to the stakeholders. This failure is mainly seen not as a kind of fraud designed to keep stakeholders from comparing companies, but rather as a lack of competency and data availability within the reporting company. Experts point to their experience that a company's carbon disclosure follows a maturity curve; consequently, first reporting is almost always incorrect, and data quality improves over time. From a stakeholder perspective, then, an implausible CO_2 inventory disclosure should not necessarily be interpreted as green-washing signal but rather as lemon signal, as the company is not able to assess its climate impact.

The role of the AUDIT STATEMENT is seen as a controversial one. While limited assurance is mostly seen as sufficient from the level of an assurance perspective, the audit scope is often seen as not necessarily suited to the purpose of identifying green-washing companies. Here, experts criticize that the audit assignment focusses on pure numbers, usually connected to scopes 1 & 2 emissions, rather than embedding pure numbers into a CLIMATE STORYLINE. This would ensure that the described climate strategy has the impact on carbon trending as verified by the auditors. Generally, experts are less interested in the CO_2 inventory number than they are in the company's strategy and how the company's value-creation process is connected with the challenges of global warming. Therefore, the TRENDING, target-setting and target achievement of carbon information should be linked to CO_2 inventory reporting.

The ability to COMPARE companies based on their reported inventory is seen as impossible, as the information disclosed by any company is far too aggregated to enable a comparison at a company level. The differentiation between efficient and non-efficient companies should be done here either by looking at trending information to judge the ability to adapt to climate-change challenges, or by looking at the product level. For a product comparison, experts rely on industry-specific KPIs and sector-specific standards.

As cross-company comparability is limited, stakeholders particularly consider the adoption of targets aligned with the global TWO-DEGREE TARGET as relevant. Due to the low number of companies which have actually set this ambitious target, reporting against this target is seen as a cherry signal. Alongside the two-degree target, honest reporting about current limitations to the climate-change agenda and potential actions on how to close these gaps are perceived as a cherry signal.

For REPORTING COMPANIES AS AGENTS that seek to send a cherry signal, the outcome of this research has MULTIPLE IMPLICATIONS. The hygienic minimum requirement for any message is to report consistently over time and across all scopes audited with limited assurance to a known standard. To convert this message into a truly green signal, this CO_2 disclosure should also be embedded into the company strategy, which includes a strategy for coping with global warming. Additionally, truly green signals feature reporting against targets aligned with the two-degree target. Honest information as to the areas in which the company is not on track against the two-degree target is perceived as a truly green signal, as this information is not associated with any green-washing strategy.

STAKEHOLDERS AS PRINCIPALS should ensure that companies fulfill the requirements of truly green signaling and actively blame companies that disclose information that does not meet the minimum requirements for reporting. The principles of consistency and completeness need to gain more attention. If the reporting company has received external assurance and failed to report according to general reporting principles, stakeholders should also communicate that the respective assurance company does not deserve the trust of society to ensure that bad reporting practices are associated with reduced public trust of the auditor. Carbon inventory is supposed to supplement a certain story a company tells about reducing its carbon emissions. To do this in a transparent manner, the reporting principles have to be applied, and the reduction effects must reflect the overall strategy that forms part of the communication. The assurance statement should not only verify total numbers but also corroborate the way the overall story combines with the numbers offered in support of that story. This ensures that the AUDITOR will not only need to confirm material changes in the inventory, but will use his or her inside view to verify that the material effects in the carbon inventory are caused by emission-reduction actions and not by other effects influencing carbon KPIs. If the audit statement does not verify both, audit companies run a significant risk of losing credibility as the assurance statements could be used to support a green-washing strategy. This could result if companies use independent effects on carbon accounting to communicate that they are green, while auditors could claim that they only reviewed the inventory according to a certain standard. A side effect of combining the green storyline and corresponding figures over the full value chain could also be an improvement in integrated reporting practices, which are currently not viewed as providing relevant information to stakeholders (Thomson, 2015; Weißenberger, 2014).

One limitation of comparing companies is the absence of sector-specific reporting standards and sector-specific KPIs. Here STANDARD-SETTERS have the ability to provide further guidance on how to compare companies and define pathways as to what KPIs have to be met by what time to be two-degree compliant. Especially the incorporation of the full value chain and focus areas could support reporting companies and stakeholders.

For experts, the empirical studies that dominate the publication landscape for the pays-to-be-green area have almost no impact. Therefore, RESEARCHERS should shift their focus to support and guide standard-setters, stakeholders and reporting companies.

II.   Environmental disclosure as measure for green performance

To test if those findings can be also confirmed in a bigger sample, the findings were operationalized and tested with three main hypotheses. The operationalization was embedded on the theoretical model and aiming to close the research gaps as shown in Figure 5.2.

The first hypothesis reviewed the basic assumption of the signaling argument by reviewing if environmental disclosure quality is associated with environmental performance. This hypothesis could be confirmed for scopes 1 & 2 performance against the disclosure quality even though the descriptive analytics showed that accounting quality is a rather poor measure to distinguish between companies with good versus bad performance. The measure of disclosure quality itself had a significant overall improvement between the reporting years 2014 and 2015. This disclosure improvement could be associated with higher media coverage of the two-degree target during the preparation period of the Paris agreement (Uni, 2015). This improvement of disclosure quality during a period of high media attention to global warming fits to the theoretical concept in which agency signals are triggered when society and regulators approach investors and companies. This triggering actually embraces the pays-to-be-green business case. It is interesting to note that the performance of the overall sample did not change significantly between 2014 and 2015 (Figure 5.4). This implies that relatively good disclosure is a poor measure for use in distinguishing between companies with good and bad performance, yet the absolute disclosure score is a poor measure for performance. This supports the theoretical concept of the institutional theory as shown in Figure 3.5 that companies align their disclosure strategy over time and copy best-in-class disclosure behaviour.

A performance check of all emissions could not be done, as the sample of companies providing a sufficient amount of complete scope 3 inventory was too small. Within the sample of companies providing sufficient information, only very few companies achieved the two-degree target over all scopes. Within the CDP sample, the self-reported emission trend does not correlate with the calculated emission trend of the CDP data set, as the correlation Table 6.4 shows. Combined with the findings on how to identify green-washing companies, these findings are a strong indicator of green-washing.

The IMPLICATIONS for the THEORETICAL CONSTRUCT are that the disclosure trend indicates that disclosure quality can be explained by legitimacy and institutional theory. The agency argumentation is also a valid theoretical construct; however, a pure carbon inventory requires supplementary information to enable a truly green signal. As most companies fail to provide sufficient information over the full value chain, the signal which is being sent can be easily copied by reporting companies. The ability to copy the respective signal could be proven in the years 2014 and 2015 in which overall disclosure quality increased significantly without any major change in performance.

From a RESEARCH PERSPECTIVE, the first implication is that the variable disclosure quality is only a poor measure to operationalize green performance (Figure 5.4). This is an important finding, as many articles within the pays-to-be-green literature use various disclosure variables to operationalize green performance. Another research implication is that the disclosure currently covers only the companies' own emissions (scopes 1 & 2) which does not cover the full value chain as requested by standards. This incomplete data limits the explanatory power of any quantitative research using emissions data. Especially the mismatch between scopes 1 & 2 performance and all scopes performance indicates that most findings of quantitative works based on emissions data have only limited informative value from an environmental perspective. This limitation is caused as scopes 1 & 2 performance is heavily influenced by vertical integration of the value-creation process. In quantitative research activities, it is extremely challenging to account for in- and outsourcing of emission-intensive activities that mainly influence the performance of scopes 1 & 2 emissions.

Meta-analytical studies indicated that operationalization for green performance and green reporting is poor. Therefore, implications of this research include the OPERATIONALIZATION OF GREEN MEASURES based on disclosure information provided to the CDP. The QUALITY was operationalized along the principles of the GHG Protocol. This is a quantitative feasible approach to enable any stakeholder or researcher to assess the disclosure quality which also meets the minimum requirements assessed in the qualitative study. The operationalization of GREEN PERFORMANCE was performed in several ways using absolute and relative emissions trends. One implication which can be used in further research is the operationalization of the guidance of compliance with the two-degree target. Meta-analytical studies indicated that a green performance measure is missing. This research gives a role model for further analysis on how to operationalize green performance in a quantitatively relevant way for own emissions and emissions over the full value chain. At the same time, this research showed that self-reported emission trends are not a good sign for carbon performance, as indicated in Tables 5.8, 5.14, and 6.4.

The most relevant implication of this research for researchers, stakeholders, reporting companies and standard-setters is the improvement of the DATA AVAILABILITY OF SCOPE 3 EMISSIONS. The quality-assurance process of assessing which companies report a complete scope 3 inventory by checking if all material has been reported were incorporated into the CDP assurance process. This assurance process aims to improve the complete data submission of companies to improve overall data availability over time. As the CDP is the most used data hub for CO_2 emissions, it is very likely that a large audience will benefit from this improvement in future.

III.   Compliance to the two-degree target

Operationalization of the two-degree target showed that the green disclosure variable is not necessary a good measure to operationalize green performance.

Figure 5.8 indicates that for scopes 1 & 2, the difference is not sufficiently large to distinguish between companies with good and poor performance. The graphic shows also that the companies do not provide sufficient information over the whole value chain and almost never reach the two-degree target on absolute emissions. This triangulation of the hypothesis as to whether disclosure can be associated with carbon performance supports the results and implications.

IV.   Assurance coverage as measure for green disclosure

The measures representing carbon disclosure and carbon performance were often perceived as poor in the literature review. As Hypothesis 1 works with a complex but effective measure to operationalize carbon disclosure, Hypothesis 3 tested whether a shortcut to the extent of using a voluntary audit could serve as an operationalization factor for scopes 1 & 2. The audit score and the disclosure score correlate highly (0.7 as shown in Table 6.4), which supports the validity of using the audit score as operationalization measure for disclosure practice. This test is performed by reviewing whether the audit score is similarly connected to carbon performance variables as the disclosure score. As shown in Table 5.15, the results are very similar, and the overall r-squared increases. For scope 3, it can be seen that these companies do not comply with the two-degree target on all scopes, no matter if the scope 3 emissions are audited or not.

The IMPLICATION that the audit score is a good measure to represent disclosure quality according to the reporting principles is that researchers or other interested stakeholders can use the audit score to operationalize green disclosure. Additionally, the increased r-squared in the analysis between carbon performance and carbon disclosure supports the theoretical setting that companies that voluntarily request assurance of their emission statement do have a positive message they would like to bring across. The check that the audit provides also represents a good way of meeting the disclosure principles expected, as the assurance statement is usually being performed against a certain disclosure standard, which usually includes the respective principles operationalized by the disclosure score. Like the disclosure score, the audit score is already a poor measure to operationalize green performance for own emission trends. The all-scope audit score cannot be associated with green performance over all scopes for absolute emission trending. The implication is that scope 3 assurance is currently no indicator for carbon performance that meets the two-degree target for the full value chain.

V.   Confirmation of control variables

One open research question from the meta-analytical review was that the question of relevant MODERATOR VARIABLES was unanswered. Studies from Guidry and Patten (2012) and Patten (2002) indicated that only company size and company sector should be used as moderator variables. Relevance of company SIZE and SECTOR for disclosure behaviour could be confirmed (Tables 6.8 and 6.9). Size, however, had no implication on the audit behaviour of the company. This could be explained either by the stronger signaling effect of the audit score compared to general disclosure or by the limited heterogeneity in company size of this data set, as the majority of CDP respondents are large, multinational, listed companies.

Additionally, according to the theoretical framework as shown in Figure 5.2, STAKEHOLDER SALIENCE also influences reporting practices. This stakeholder salience was operationalized with the multiple variables representing the legal setting, including the effectiveness of the government and the company headquarters location and the influence of the media as well as the energy intensity of the industry and the exposure to investors in indices. In sum, these variables mainly have a significant influence on the disclosure behaviour of the company, which confirms the mediating effect of stakeholder power. The significance of those mediating effects gets partially lost by slicing the data set based on disclosure behaviour which is an indicator of the weakness of this mediating effect (Table 5.20). The IMPLICATION of this finding is that, in future, carbon accounting researchers should always use size and sector as moderator variables and should assess which further mediating effects shall be taken into account.

VI.   Agency theory as argumentation for CO_2 disclosure

The MIXED-METHOD RESEARCH QUESTION as to whether the ancillary conditions for signaling in the field of CO_2 disclosure are fulfilled, so that signaling can explain high-quality disclosure beyond the minimum expectations of stakeholder pressure, cannot be answered fully in the affirmative. Positive indicators for the research question include the positive relationship between disclosure practice, especially the use of audit services, and the emission trend for own emissions. If these indicators were viewed in isolation, the interpretation would be a confirmation that signaling can explain high-quality voluntary disclosure. This is especially well underpinned, as it is consistent with the typically positive self-description of the reporting company's green strategy. This fits with the finding from the interviews that the CO_2 emission trend should be seen as supplementary material to explain the overall green actions. Voluntary disclosure of detailed carbon-footprint information should be seen as evidence that a company is achieving the performance communicated. From a theoretical perspective, the signaling effect is also a fitting theoretical backing for high-quality reporting.

Contradicting this interpretation is that the side condition that the signal cannot be easily copied is not sufficiently met. Starting within the performance of the own emissions (Figure 5.7), it turns out that a significant share of the companies that fail to meet the two-degree target for own emissions rank among the leading firms in terms of disclosure practice. At the same time, about 50% of the group of non-leaders consists of companies that achieve compliance in terms of scopes 1 & 2 inventory. This shows that reporting practice and the disclosure performance are not a sufficient element for use in distinguishing between green and non-green companies. Incorporation into the green strategy also seems to be hard to prove, as experts indicated that only very few companies incorporate green actions into their strategy , while a significant share of companies claim to be green.

A stronger indicator that companies currently fail to send signals which cannot be copied is the incorporation of the full value chain. Hypothesis 2 showed the small total number of companies that successfully manage their carbon inventory over all scopes and disclose this properly as recommended in the CDP and GRI standard. As Figure 5.7 shows, only a single-digit percentage of companies manage their carbon footprint over the whole value chain. Comparing this sample to the entire population of companies worldwide, this percentage is probably even lower, as only a limited number of companies respond to the CDP questionnaire. Additionally, it is likely that this sample is biased to companies with a higher awareness for environmental topics, as column Abs_All represents only companies that submitted a full carbon inventory to the CDP. This includes all material categories for scope 3. Figure 5.8 illustrates the absolute numbers of companies that are two-degree compliant across all scopes for the years 2014 & 2015. This quantitative assessment shows that if the full value chain reporting is required to send a truly green signal to the market, the vast majority of companies chooses to not send this signal. The very few companies which send the respective true signal, however, fail to send a carbon-efficient signal over time, as almost none of them is able to meet the required CO_2 reduction to be compliant to the two-degree target. Therefore, the underlying mixed-method research question has to be neglected as companies fail to send green signals which enable stakeholders to distinguish between cherries and lemons. An example for green signals which can be copied easily are disclosed information, which include only own emissions, as competitors with an outsourcing strategy can easily achieve similar emission-reduction trends. Or companies fail to send truly green signals over their full value chain. Here, in principle it is possible that companies send a positive signal over the full value chain; however, the companies within the sample are so few that their existence could be also explained by random performance distribution of the companies in scope and not necessarily grounded on a theoretical disclosure construct.

Accordingly IMPLICATIONS are stakeholders cannot rely on the amount of information disclosed and treat this as an indicator of green performance as companies will combine a GREEN-WASHING STRATEGY with an outsourcing strategy for carbon-intense activities. This combination enables companies to gain free green rents. This is consistent with expert interview findings indicating that carbon results should preferably be used in a qualitative rather than a quantitative setting. For this reason, there is a strong demand for stricter regulations on the part of the expert community. The target of these regulations must be to force companies to adopt actions that move society towards carbon-neutrality. The two-degree plan requires this by no later than 2050. This explains the dominance of the legitimacy arguments in current research.

The ECONOMIC IMPLICATION references the second mixed-method research question. As long as companies fail to provide evidence that carbon emissions are reduced over the full value chain, stakeholders cannot distinguish between carbon-efficient and carbon non-efficient companies. Companies could green-wash their carbon inventory by outsourcing carbon-intense activities. This leads to a positive CO_2 trend. Stakeholders thus have no incentive to provide green rents based on companies' disclosure behavior. This might explain the inconsistent link in metastudies measuring the pays-to-be-green effect by using disclosure variables for green performance.

Assuming that voluntary actions of companies are a core pillar to fight climate change; assuming that these voluntary actions need to be valued by the stakeholder community; and assuming that this community uses voluntary disclosure to assess whether companies are acting in support of reductions in global-warming potential, the negation of the ancillary conditions has an ecological implication as well. This ecological implication is reflected in current global warming effects. Companies that fail to disclose their total global warming effect across all scopes fail to enable stakeholders to judge if the company complies with any two-degree scenario. This can explain the current mismatch between the positive communication of companies on how well they reduce their impact on climate change and the absolute increase in global emissions.

This mismatch is consistent with disclosure practices, and with criticism by experts that companies disclose and manage only their own emissions while ignoring the rest of the value chain.

B.   Limitations

The limitations of the results shown in the QUALITATIVE STUDY are mainly connected to the sampled experts. Starting with the sampling size of eleven, the respective results are not representative. Additionally, the sampled experts were all sampled from communities which are interested in fighting climate change and had a positive view of green disclosure. Reporting experts which do not believe in the human impact on global warming or persons which do not see the need to fight climate change were not interviewed. Therefore, opinions from experts who admitted the usage of green-washing strategies were missing. Next to the experts from companies which used a green-washing strategy also stakeholders which are choosing a green-washing strategy were missing: e.g. institutional investors which signed the CDP with the purpose of communicating to their stakeholders that they actively fight climate change but are not interested in this topic at all. Another limitation based on the sample is the theoretical construct and the application of the institutional theory. As the majority of experts are either involved in standard-setting processes or serve as representatives of institutions that foster comparability of carbon disclosure, it was expected that indicators of institutional effects would be confirmed. As the expert interviews were sampled from heterogeneous backgrounds, the interviews themselves could only be held in a semi-structured manner to learn from the unique fields of specialization. E.g. an NGO working on standard-setting for carbon-disclosure standards has a background completely different to that of a partner in an investment company specialising in green products. This limits the ability to compare answers and cluster answers into groups. The reproducibility of the interviews was limited in two ways. First, the interviewer identified himself during the interview as a subject-matter expert to increase the information given during the interview. If the interview had been reproduced, the answers could differ. Second, some experts refused recording of the interview, that limits the reproducibility to the minutes taken. In terms of content, the green signaling was focusing on carbon disclosure, which is only one of many indicators that can be used to send green signals to the market.

The results of the QUANTITATIVE STUDY are also associated with several limitations. The first limitation is associated with the SAMPLE used. We start with the sample and the underlying limitations on representativeness. The sample was taken as the whole dataset, excluding financial institutions, submitted voluntarily to the CDP. As the CDP is sponsored by institutional investors, the sample is biased by large multinational companies which are stock-listed and owned (partially) by institutional investors. Family-owned businesses and/or SMEs are mainly missing. Additionally, the signaling was reduced to the data self-submitted to the CDP. A consistency check against other sources, e.g. published on websites, CSR reports, annual reports, reporting schemes or external estimations, was not performed.

Other limitations are associated with the OPERATIONALIZATION of the different elements. For disclosure quality, the reporting principles were mapped to questions from the CDP. A triangulation on how well those questions actually represent the underlying variables was not performed. The corresponding audit score was also one part of the disclosure score. The correlation (Table 6.4) between the disclosure score and the audit score could therefore also partially be explained by overlap in the calculation. CO_2 performance was only measured with multiple questions from the CDP as well as the absolute and relative performance trend. The effect of outsourcing activities over time could be only covered by the financial KPI value add. The extent to which emission-intensive steps were in- or outsourced especially frequently was not measured. Missing operationalization measures refer to sector-specific KPIs at the product level, indicators for eco innovation, the ability to incorporate disruptive technologies without any emissions, the current level of intensity based on industry-specific production KPIs and the ability of the company to benefit from products that actually reduce the global warming impact.

The operationalization of MODERATOR and MEDIATOR effects was based upon previous research by Guidry and Patten (2012) and Guenther et al. (2015). Factors for stakeholder influence were always operationalized for the country of headquarters or the industry sector with the highest revenue share. Additionally to this aggregation on a singly industry and a single country the question as to how well these variables represent stakeholder salience is open. The model did not operationalize the different strategies concerning how green the stakeholders of the company actually are. This limits the results, as the willingness of stakeholders to provide green rents or penalize non-green companies triggers the requirement of companies to send corresponding green signals. With regard to penalties, historic environmental failures of companies were not operationalized as mediating effect, even though researchers showed that disclosure improves after a negative event.

From a data-analytical perspective, a deep dive into a given sector is missing. Also, the whole analytical part is not able to assess causality. The respective time trending analyses were not performed. Even though the qualitative results highlighted the importance of scope 3, the analysis over the full value chain was only performed based on descriptive information, as data availability was not sufficient.

C. Future research prospects

As result of this research, several research gaps were identified. One area of the open research gap is associated with PROXIES representing green performance. CO_2 scope 1 divided by revenue is regularly used as a proxy in the investment community to operationalize CO_2 efficiency. This is not sufficient, as this KPI can be easily manipulated by outsourcing emission-intensive activities, reducing scope 1 emissions without any environmental impact. Developing sector-specific KPIs could also help identify sector classifications that reflect company similarities in the field of CO_2 reporting. In the field of efficiency KPIs, the sector-specific assessment if a company is two-degree compliant is especially interesting. This research gap could be reviewed from an inventory or product perspective. Currently, stakeholders are dependent on companies' internal scientific targets. Therefore, the academic community needs to support the breakdown of the macroeconomic data to include the implications of the reporting company for the two-degree scenario, and translate these implications to KPIs that can be externally calculated. KPIs that can be used to track a company's status towards the two-degree target are currently missing (Walls et al., 2011).

In the field of KPIs, the research showed that the availability of full value chain reports is very limited. Studies showing how to easily set up a full value chain reporting which can be improved over time and based on the focus of material aspects could support the availability of this information. The complete inventory is connected to transparent information. This includes transparency as to how the data were gathered, as defined in the GHGP transparency principle. The common practice of naming a list of different methods is not sufficient. There must be an understanding of what proportions of the inventory are calculated based on which methods. This enables stakeholders to judge whether the company is actually in a position to manage its material inventory, and understand how accurate the respective inventory can be under the best circumstances. Applying these principles consistently over time would increase stakeholders' ability to make judgements based on the carbon inventory trend. Researchers could support the journey by showing standard-setters', auditors' and reporting companies ways of disclosing and verifying this information.

In connection to this honest reporting, research questions arise as to why companies are failing to send truly green signals concerning carbon-inventory reporting. As experts indicated that carbon inventory reporting should be seen as supplementary information, the question arises as to how to send truly green signals that cannot be copied. Remaining in the agency theory, the question also arises as to how to increase the cost of incorrect signals by penalizing green-washing signals and the information intermediates that underpin these green-washing signals.

Bibliography

(2004). The Greenhouse Gas Protocol - A Corporate Accounting and Reporting Standard. Technical report, World Business Council for Sustainable Development; World Resource Institute, Conches-Geneva.

(2008). AA1000 Assurance Standard.

(2010). GHG Protocol Corporate Value Chain (Scope 3) and Product Life Cycle Standards. Technical Report Scope 3, WBCSD, WRI.

(2011a). Corporate Value Chain (Scope 3) Accounting and Reporting Standard. Technical report, WBCSD, WRI, Washington.

(2011b). Guidance for Calculating Scope 3 Emissions. Technical Report August, WBCSD, WRI.

(2011). ISAE 3000 - Assurance Engagements Other Than Audits or Reviews of Historical Financial Information. Technical report, IAASB.

(2011). *ISAE 3410, Assurance Engagements on Greenhouse Gas Statements*. Number January. International Auditing and Assurance Standards Board.

(2011c). Overview of GHG Protocol scopes and emissions across the value chain. Technical report, WBCSD; World Resources Institute.

(2013a). Basis for conclusions - International IR Framework.

(2013). *G4 Sustainability Reporting Guidelines*. Global Reporting Initiatives, Amsterdam, version 3. edition.

(2013b). IIRC Pilot Programme Yearbook 2013.

(2013c). London senkt die CO2-Grenzwerte für Autos deutlich ab.

(2013). *The Green Investment Report: The ways and means to unlock private finance for green growth*. World Economic Forum.

(2014a). CDP DACH Report. Technical report, CDP.

(2014b). CDP's 2014 Climate Change Information Request. Technical Report 1122330, CDP, GHGP, WRI.

(2014a). Get to grips with the six capitals.

(2014b). Realizing the benefits: The impact of Integrated Reporting. Technical Report www.blacksunplc.com; www.theiirc.org, IIRC, Black Sun.

(2014). Status of Ratification of the Kyoto Protocol. Technical report, United Nations.

(2014). The integrated report. Technical Report June, The Institute of Directors in Southern Africa.

(2015a). About The Climate Registry.

(2015). Adoption of the Paris Agreement. Technical Report December, United Nations, Paris.

(2015). Carbon Asset Risk: Discussion Framework. Technical report, World Resource Institute.

(2015b). CDP 2014 Climate Change Scoring Methodology. Technical report, CDP.

(2015a). CDP Academics. Technical report, CDP.

(2015b). CDP Climate Change Report 2015 - DACH 350+ Edition. Technical report, CDP.

(2015c). CDP Climate Change Report 2015: The mainstreaming of low-carbon on Wall Street US edition based on the S&P 500 Index. Technical Report November, CDP.

(2015d). CDP Policy briefing - Business ad the Paris agreement. Technical report, CDP, Paris.

(2015e). CDP Reporting Roadmap Climate Change. Technical report, CDP.

(2015f). CDP's 2016 Climate Change Information Request. (1122330):1–17.

(2015a). CDSB Board Members.

(2015b). CDSB Framework. Technical Report June, CDSB.

(2015). Climate Strategies and Metrics: Exploring Options for Institutional Investors. Technical report, UNEP-FI, WRI, 2 degree Investing Initiative.

(2015). IIRC Partners.

(2015a). MSCI KLD 400 Social Index Methodology.

(2015b). MSCI USA IMI Sector Indexes.

(2015). Responsible Corporate Engagement in Climate Policy - over 100 companies committed to action. Technical Report December, Caring for Climate, UN Global Compact, UNEP, UNCCC, WRI, CDP, WWF, PRI, Ceres, The climate group.

(2015g). Sectoral Decarbonization Approach (SDA) - A product of the science based Targets initiative. Technical Report May, CDP, WRI, WWF.

(2015). The business case for responsible corporate adaption: Strengthening Private Sector and Community Resilience. Technical report, UN Global Compact, UNFCCC, UNEP, CDP, The CEO Water Mandate, Oxfam, Rainforest Alliance, ARISE, University of Notre Dame, WRI.

(2016a). Climate disclosure makes sound business sense. Technical report, CDP, London.

(2016). Commit to setting science based targets.

(2016b). Guidance for companies reporting on climate change on behalf of investors & supply chain members 2016. Technical report, CDP.

(2016). Investor climate disclosure. Technical Report May, 2degrees investing initiative, Paris, New York, Berlin, London.

(2016). MSCI low carbon indexes family.

(2016). Science based targets - case studies. Technical report, WRI.

(2016). The Rockefeller Family Fund Decision to Divest.

(2016c). What third party verification standards are appropriate when reporting to CDP's climate change program? Technical report, CDP.

(2017). CDSB Start page.

(2017). Companies taking action. Technical report, We mean business Coalition.

(2017). MSCI ACWI Sustainable Impact Index.

Aguilera-Caracuel, J. and Ortiz-de Mandojana, N. (2013). Green Innovation and Financial Performance: An Institutional Approach. *Organization & Environment*, 26(4):365–385.

Aguinis, H. and Glavas, A. (2012). What We Know and Don't Know About Corporate Social Responsibility: A Review and Research Agenda. *Journal of Management*.

Akerlof, G. A. (1970). The Market for "Lemons": Quality Uncertainty and the Market Mechanism. *The Quarterly Journal of Economics*, 84(3):488–500.

Albertini, E. (2013). Does Environmental Management Improve Financial Performance? A Meta-Analytical Review. *Organization & Environment*, 26(4):431–457.

Allianz S.E. (2015). Allianz Statement on Coal - based Investments.

Allison, C., Collins, M., and Fisher, K. (2012). GHG Protocol Product Life Cycle Accounting and Reporting Standard Sector Guidance for Pharmaceutical and Medical Device Products Consultation Draft June 2012 GHG Protocol Product Life Cycle Accounting and Reporting Standard Sector Guidance for Pharmaceuti. Technical Report June, GHGP.

Andrew, J. and Cortese, C. (2011a). Accounting for climate change and the self-regulation of carbon disclosures. *Accounting Forum*, 35(3):130–138.

Andrew, J. and Cortese, C. (2011b). Carbon Disclosures: Comparability, the Carbon Disclosure Project and the Greenhouse Gas Protocol. *Australian Accounting, Business and Finance Journal*, 5(4):6–18.

Andrew, J. and Cortese, C. (2013). Free market environmentalism and the neoliberal project: The case of the Climate Disclosure Standards Board. *Critical Perspectives on Accounting*, 24(6):397–409.

Aphalo, P. J. (2016). *Learn R ...as you learnt your mother tongue*. Leanpub, Helsinki.

Appel, F., Beckmann, C., Hoeffe, O., Tomoff, K., Lohmeier, M., Ellermann, D., Sprinz, D., Robert, S., Zhou, R., Bolton, L., van der Berg, S., Kiwitt, P., Straube, F., Doch, S., and Wend, R. (2010). *Delivering Tomorrow Towards Sustainable Logistics How Business Innovation and Green Demand Drive a Carbon Efficient Industry*. Deutsche Post Headquarters represented by Dr Christof Erhard, Bonn, 1. auflage edition.

Arbeitskreis Externe Unternehmensrechnung der Schmalenbach-Gesellschaft für Betriebswirtschaft (2015). Nichtfinanzielle Leistungsindikatoren - Bedeutung für die Finanzberichterstattung. *ZFBF*, pages 235–258.

Arena, C., Bozzolan, S., and Michelon, G. (2014). Environmental Reporting: Transparency to Stakeholders or Stakeholder Manipulation? An Analysis of Disclosure Tone and the Role of the Board of Directors. *Corporate Social Responsibility and Environmental Management*, 361(May 2014):n/a–n/a.

Armstrong, E. (2011). Voluntary greenhouse gas reporting. *Environmental Quality Management*, 20:29–42.

Arnold, J. (2011). *Die Kommunikation gesellschaftlicher Verantwortung am nachhaltigen Kapitalmarkt*. VS Verlag, Hohenheim.

Ascui, F. (2014). A Review of Carbon Accounting in the Social and Environmental Accounting Literature: What Can it Contribute to the Debate? *Social and Environmental Accountability Journal*, 34(1):6–28.

Ascui, F. and Lovell, H. (2011). As frames collide: making sense of carbon accounting. *Accounting, Auditing & Accountability Journal*, 24(8):978–999.

Atteridge, A., Siebert, C. K., Klein, R. J. T., Butler, C., and Tella, P. (2009). *Bilateral finance institutions and climate change: A mapping of climate portfolios*, volume 2010.

Auerbach, A. and Cutler, D. (2015). Carbon Taxes II.

Auvinen, H., Mäkelä, K., Johanson, B., and Ruesch, M. (2011a). Methodologies for emission calculations - Best practices , implications and future needs. page 28.

Auvinen, H., Mäkelä, K., Lischke, A., Burmeister, A., de Ree, D., and Ton, J. (2011b). D 2.1 COFRET - Existing methods and tools for calculation of carbon footprint of transport and logistics. Technical Report deliverable 2.1, VTT Technical Research Centre of Finland.

Baker, C. (2016). Emission impossible? Technical Report 5.

Balderjahn, I. (2013). Nachhaltigkeit aus Konsumentensicht - Nachhaltiger Konsum -. In *Schmalenbachtungung*, number April, pages 1–25, Köln.

Bansal, P. (2005). Evolving sustainably: A longitudinal study of corporate sustainable development. *Strategic Management Journal*, 26(3):197–218.

Bansal, P. and Roth, K. (2000). Why Companies Go Green : Responsiveness. *Academy of Management*, 43(4):717–736.

Barkemeyer, R., Comyns, B., Figge, F., and Napolitano, G. (2014). CEO statements in sustainability reports: Substantive information or background noise? *Accounting Forum*, 38(4):241–257.

Barth, M. E. and McNichols, M. F. (1994). Estimation and Market Valuation of Environmental Liabilities Relating to Superfund Sites. *Journal of Accounting Research*, 32(3):177–209.

Barton, A. H. and Lazarsfeld, P. F. (1979). Einige Funktionen von qualitativer Analyse in der Sozialforschung. pages 41–89.

Basacik, L., Kutner, M., Buck, B., Dreyfus, R., Espinach, L., Hagen, S., and Kriege, K. (2015). Linking GRI and CDP: Water. Technical report, Global Reporting Initiative, CDP.

based targets, S. (2016). Science based targets - methods.

Beaver, W. H. (1981). Market Efficiency. *The Accounting Review*, 56(1):23–37.

Bebbington, J. and Larrinaga-González, C. (2008). Carbon Trading: Accounting and Reporting Issues.

Bednar, M. K. and Westphal, J. D. (2007). Research Methodology in Strategy and Management Emerald Book Chapter : Surveying the Corporate Elite : Theoretical and Practical Guidance on Improving Response Rates and Response Quality in Top Management Survey Questionnaires SURVEYING THE CORPORATE ELIT. *RESEARCH METHODOLOGY IN STRATEGY AND MANAGEMENT*, 1(1).

Bennett, M., Schaltegger, S., and Zvezdov, D. (2013). *Exploring Corporate Practices in Management Accounting for Sustainability*. London.

Bergius, S. (2015). Ein Jahr hat viel bewegt. *Nachhaltige Investments*, (12).

Berthelot, S. and Robert, A.-M. (2011). Climate change disclosures: An examination of Canadian oil and gas firms. *Issues in Social & Environmental Accounting*, 5(1):106–123.

Best, P. J., Buckby, S., and Tan, C. (2001). Evidence of the Audit Expectation Gap in Singapore. *Managerial Auditing Journal*, 16(3):133–144.

Blacconiere, W. G. and Patten, D. M. (1994). Environmental disclosures, regulatory costs, and changes in firm value. *Journal of Accounting and Economics*, 18(3):357–377.

Boesso, G. and Kumar, K. (2007). Drivers of corporate voluntary disclosure: A framework and empirical evidence from Italy and the United States. *Accounting, Auditing & Accountability Journal*, 20(2):269–296.

Borkowski, S. C., Welsh, M., and Wentzel, K. (2011). Sustainability reporting assurance: your next growth area? *CPA Journal*, 82(2):24–27.

Botta, J., Freigang, S., Hufschlag, K., Spittler, S., and Weber, J. (2012). *Carbon Accounting und Controlling Grundlagen und Praxisbeispiel Deutsche Post DHL.* Number 83. Wiley, Weinheim, 1. auflage edition.

Bowen, F. and Wittneben, B. (2011). Carbon accounting - Negotiating accuracy, consistency and certainty across organisational field. *Accounting, Auditing & Accountability Journal,* 24(8):1022–1036.

Bragdon, J. H. and a.T Marlin, J. (1972). Is pollution profitable? *Risk Management,* April:9–18.

Brouhle, K. and Harrington, D. R. (2010). GHG Registries: Participation and Performance Under the Canadian Voluntary Climate Challenge Program. *Environmental and Resource Economics,* 47(4):521–548.

Brown, J. and Dillard, J. (2014). Integrated reporting: On the need for broadening out and opening up. *Accounting, Auditing & Accountability Journal,* 27(7):1120–1156.

Buchholtz, A. K. and Carroll, A. B. (2012). *Business & society: Ethics & stakeholder management.* South-Western Cengage Learning.

Buchner, B., Falconer, A., Hervé-Mignucci, M., Trabacchi, C., and Brinkman, M. (2011). The Landscape of Climate Finance: A CPI Report. Technical report.

Bundesministerium der Justiz (2012). Deutscher Rechnungslegungs Standard Nr 20 (DRS 20) Konzernlagebericht.

Bundesumweltamt (2013). Luft - Regelungen und Strategien.

Burck, J. and Marten, F. (2016). Questionnaire Climate Change Performance Index 2017.

Burritt, R. L., Schaltegger, S., and Zvezdov, D. (2011). Carbon Management Accounting: Explaining Practice in Leading German Companies. *Australian Accounting Review,* 21(1):80–98.

Burton, C. D. (2010). An inconvient risk: Climate change disclosure and the burden of corporations. *Administrative Law Review,* 62(4):1287–1305.

Busch, T. and Hoffmann, V. H. (2011). How Hot Is Your Bottom Line? Linking Carbon and Financial Performance. *Business & Society,* 50(2):233–265.

Caffo, B. (2015). Developing Data Products in R. *R Software,* page 52.

Calza, F., Profumo, G., and Tutore, I. (2014). Corporate Ownership and Environmental Proactivity. *Business Strategy and the Environment,* pages n/a–n/a.

Campbell, J. L. (2007). Why would corporations behave in socially responsible ways? An institutional theory of corporate social responsibility. *Academy of Management Review,* 32(3):946–967.

Carnegie Mellon University (2016). Economic Input-Output Life Cycle Assessment - Carnegie Mellon University.

Carroll, A. B. (1979). Three-Dimensional Conceptual Model of Corporate Performance. *Academy of Management Journal,* 4(4):497–505.

Carroll, A. B. (1999). Corporate Social Responsibility: Evolution of a Definitional Construct. *Business & Society,* 38(3):268–295.

Carroll, A. B. and Shabana, K. M. (2010). The Business Case for Corporate Social Responsibility: A Review of Concepts, Research and Practice. *International Journal of Management Reviews,* 12(1):85–105.

Carroll, A. B. and Shabana, K. M. (2011). The Business Case for Corporate Social Responsibility. *Director Notes*, (June):1–7.

Castka, P. and Balzarova, M. a. (2008). ISO 26000 and supply chains - On the diffusion of the social responsibility standard. *International Journal of Production Economics*, 111(2):274–286.

Cavana, R. Y. and Becker, C. S. (2016). An exploratory analysis of legitmation strategies used in sustainability reporting of negative incidents. In *Conference proceedings EURAM*, Paris.

CDP (2015h). CDP Clean & Complete Dataset - Reference Methodology for Companies.

CDP (2015i). CDP 's 2015 Climate Change Information Request.

CDP, Bureau Veritas, LRQA Lloyd's Register Quality Assurance, PwC, and TÜV Nord (2011). Carbon Disclosure Project Verification of climate data. Technical report.

ChemCologne (2016). The Verbund system - Factor of success for the region.

Cheng, M., Green, W., Conradie, P., Konishi, N., and Romi, A. (2014). The International Integrated Reporting Framework: Key Issues and Future Research Opportunities. *Journal of International Financial Management & Accounting*, 25(1):90–120.

Cheynel, E. (2013). A theory of voluntary disclosure and cost of capital. *Review of Accounting Studies*, 18(4):987–1020.

Chi, W., Dhaliwal, D., Li, O. Z., and Lin, T.-H. (2013). Voluntary Reporting Incentives and Reporting Quality: Evidence from A Reporting Regime Change for Private Firms in Taiwan. *Contemporary Accounting Research*, 30(4):1462–1489.

Chithambo, L. and Tauringana, V. (2014). Company specific determinants of greenhouse gases disclosures. *Journal of Applied Accounting Research*, 15(3):323–338.

Cho, C. H., Guidry, R. P., Hageman, A. M., and Patten, D. M. (2012). Do actions speak louder than words? An empirical investigation of corporate environmental reputation. *Accounting, Organizations and Society*, 37(1):14–25.

Cho, C. H., Michelon, G., Patten, D. M., and Roberts, R. W. (2014). CSR report assurance in the USA: an empirical investigation of determinants and effects. *International Journal of Distribuition & Logistics Management*, 5(2):130–148.

Cho, C. H. and Patten, D. M. (2007). The role of environmental disclosures as tools of legitimacy: A research note. *Accounting, Organizations and Society*, 32(7-8):639–647.

Cho, C. H. and Roberts, R. W. (2010). International Journal of Accounting Information Systems Environmental reporting on the internet by America's Toxic 100: Legitimacy and self-presentation. *International Journal of Accounting Information Systems*, 11(1):1–16.

Clark, G., Feiner, A., and Viehs, M. (2014). Stockholder to Stakeholder. *SSRN 2508281*.

Clark, G. L., Feiner, A., and Viehs, M. (2015). From the Stockholder to the Stakeholder. Technical Report March.

Clarkson, P. M., Li, Y., Richardson, G. D., and Vasvari, F. P. (2008). Revisiting the relation between environmental performance and environmental disclosure : An empirical analysis. *Accounting, Organizations and Society*, 33:303–327.

Clarkson, P. M., Li, Y., Richardson, G. D., and Vasvari, F. P. (2011). Does it really pay to be green? Determinants and consequences of proactive environmental strategies. *Journal of Accounting and Public Policy*, 30:122–144.

CMU (2013). Assumptions, Uncertainty, and other Considerations with the EIO-LCA Method.

Cochran, I., Eschalier, C., and Deheza, M. (2015). Lessons from the use of climate-related decision- making standards and tools by DFIs to facilitate the transition to a low-carbon, climate-resilient future. Technical report, I4CE - Institute for Climate Economics, Paris.

Comyns, B. and Figge, F. (2015). Greenhouse gas reporting quality in the oil and gas industry. *Accounting, Auditing & Accountability Journal*, 28(3):403–433.

Comyns, B., Figge, F., Hahn, T., and Barkemeyer, R. (2013). Sustainability reporting: The role of 'Search', 'Experience' and 'Credence' information. *Accounting Forum*, 37(3):231–243.

Connelly, B. L., Certo, T. S., Ireland, D. R., and Reutzel, C. R. (2011). Signaling Theory: A Review and Assessment. *Journal of Management*, 37(1):39–67.

Cooper, S. M. and Owen, D. L. (2007). Corporate social reporting and stakeholder account-ability: The missing link. *Accounting, Organizations and Society*, 32(7-8):649–667.

Cormier, D., Magnan, M., and Van Velthoven, B. (2005). Environmental disclosure quality in large German companies: Economic incentives, public pressures or institutional conditions? *European Accounting Review*, 14(1):3–39.

Cotter, J. and Najah, M. M. (2012). Institutional investor influence on global climate change disclosure practices. *Australian Journal of Management*, 37(2):169–187.

Crawley, M. J. (2012). *Statistik mit R*. Wiley, Weinheim, 1. auflage edition.

Creswell, J. W. (2014). *Research design: qualitative, quantitative and mixed methods approaches*. 4 edition.

Creswell, J. W. and Clark, P. V. (2011). *Designing and conducting mixed-methods research*. Sage, 2. edition.

Croissant, Y., Millo, G., Tappe, K., Toomet, O., Kleiber, C., Zeileis, A., Henningsen, A., Andronic, L., and Schoenfelder, N. (2016). Linear Models for Panel Data Depends.

Cruz, J. M. and Matsypura, D. (2009). Supply chain networks with corporate social responsibility through integrated environmental decision-making. *International Journal of Production Research*, 47(3):621–648.

Dahl, D. B. (2016). Export Tables to LaTeX or HTML.

Dahlsrud, A. (2008). How Corporate Social Responsibility is Defined: an Analysis of 37 De nitions. *Corporate Social Responsibility and Environmental Management*, 13(November 2006):1–13.

Daróczi, G. and Tsegelskyi, R. (2016). An R Pandoc Writer.

Dechant, K. and Altman, B. (1994). Environmental leadership: From compliance to compet-itive advantage. *Academy of Management Perspectives*, 8(3):7–20.

Deegan, C. and Rankin, M. (1997). The materiality of environmental information to users of annual reports. *Accounting, Auditing & Accountability Journal*, 10(4):562–583.

Deegan, C., Rankin, M., and Tobin, J. (2002). *An examination of the corporate social and environmental disclosures of BHP from 1983-1997: A test of legitimacy theory*, volume 15.

Delmas, M., Etzion, D., and Nairn-Birch, N. (2013). Triangulating environmental performanc: What do corporate social responsibility ratings really capture? *Academy of Management Perspectives*, 27:255–267.

Department for Environment Food & Rural Affairs (2014). Waste legislation and regulations.

Depoers, F., Jeanjean, T., and Jérôme, T. (2014). Voluntary Disclosure of Greenhouse Gas Emissions: Contrasting the Carbon Disclosure Project and Corporate Reports. *Journal of Business Ethics.*

Deutsche Energie Agentur (2006). Plattform Biogaspartner: Kißlegg-Rahmhaus.

Dhaliwal, D. S., Li, O. Z., Tsang, A., and Yang, Y. G. (2009). Voluntary Non-Financial Disclosure and the Cost of Equity Capital: The Case of Corporate Social Responsibility Reporting. *SSRN Electronic Journal.*

Dhaliwal, D. S., Li, O. Z., Tsang, A., and Yang, Y. G. (2011). Voluntary nonfinancial disclosure and the cost of equity capital: The initiation of corporate social responsibility reporting. *Accounting Review*, 86(1):59–100.

Dhaliwal, D. S., Radhakrishnan, S., Tsang, A., and Yang, Y. G. (2012). Nonfinancial disclosure and analyst forecast accuracy: International evidence on corporate social responsibility disclosure. *Accounting Review*, 87(3):723–759.

Die Deutsche Bundesregierung (2015). Energiewende.

Dienes, D., Sassen, R., and Fischer, J. (2016). What are the drivers of sustainability reporting? A systematic review. *Sustainability Accounting, Management and Policy Journal*, 2(7).

Dilling, P. F. a. (2010). Sustainability Reporting In A Global Context: What Are The Characteristics Of Corporations That Provide High Quality Sustainability Reports - An Empirical Analysis. *The International Business & Economics Research Journal*, 9(1):19–30.

DiMaggio, P. J. and Powell, W. W. (1983). The Iron Cage Revisited: Institutional Isomorphism and Collective Rationality in Organizational Fields. *American Sociological Review*, 48(2):147–159.

DIN Deutsches Institut für Normung (2009). DIN EN ISO 14001.

DIN Deutsches Institut für Normung (2011). DIN EN ISO 50001.

Dixon-Fowler, H. R., Slater, D. J., Johnson, J. L., Ellstrand, A. E., and Romi, A. M. (2013). Beyond "Does it Pay to be Green?" A Meta-Analysis of Moderators of the CEP-CFP Relationship.

Doda, B., Gennaioli, C., Gouldson, A., Grover, D., and Sullivan, R. (2015). Are Corporate Carbon Management Practices Reducing Corporate Carbon Emissions? *Corporate Social Responsibility and Environmental Management*, (September):n/a–n/a.

Dohrn, M. (2004). *Entscheidungsrelevanz des Fair Value-Accounting am Beispiel von IAS 39 und IAS 40.* Baetge, Jörg; Kirsch, Hans-Jürgen; Thiele, Stefan, Lohmar Köln, rechnungsl edition.

Dolic, D. (2004). *Statistik mit R.* Oldenburg Verlag, Wien, München.

Donaldson, T. and Preston, L. (1995). The Stakeholder Theory of the Corporation: Concepts , Evidence , and Implications. *The Academy of Management Review*, 20(1):65–91.

Doran, K. L. and Quinn, E. L. (2009). Climate change risk disclosure: a sector by sector analysis of sec 10-k filings from 1995-2008. *North Carolina Journal of International Law and Commercial Regulation*, 34:722–766.

Downie, J. and Stubbs, W. (2011). Evaluation of Australian companies' scope 3 greenhouse gas emissions assessments. *Journal of Cleaner Production*, pages 1–8.

Downie, J. and Stubbs, W. (2012). Corporate carbon strategies and greenhouse gas emission assessments: The implications of scope 3 emission factor selection. *Business Strategy and the Environment*, 21(6):412–422.

DPA (2013). Schifffahrt Neue Umweltvorschriften treiben Kosten in die Höhe.

Dragomir, V. D. (2012). The disclosure of industrial greenhouse gas emissions: a critical assessment of corporate sustainability reports. *Journal of Cleaner Production*, 29-30:222–237.

D'Souza, J. D., Baginski, S., Beneish, D., Biddle, G., Elliott, J., Hopkins, P., Johnson, M., Maines, L., Connor, K. O., Pae, S., Pratt, J., Salamon, J., Shevlin, T., Skinner, D., Smith, A., Sprinkle, G., Stanford, M., Swieringa, B., Tasker, S., and Wahlen, J. (2004). Voluntary Disclosure in a Multi-Audience Setting: An Empirical Investigation. *American Accounting Association*, 79(4):921–947.

Dye, R. A. (1986). Proprietary and Nonproprietary Disclosures. *Journal of Business*, 59(2):331–366.

Eccles, R. G. and Saltzman, D. (2011). Achieving Sustainability Through Integrated Reporting. *Stanford Social Innovation Review*, Summer:56–61.

Eccles, R. G. and Serafeim, G. (2012). The impact of corporate sustainability on organizational processes and and performance. *National Bureau of economic research*.

Eckmueller, B. (2013). Carbon Disclosure Leadership index: Steria wird klassenbester IT Dienstleister.

ECOpoint Inc. (2015). Emission Standards: United States.

Eleftheriadis, I. M. and Anagnostopoulou, E. G. (2014). Relationship between Corporate Climate Change Disclosures and Firm Factors. *Business Strategy and the Environment*, pages 1–10.

Elkington, J. (1997). Cannibals with forks. *The triple bottom line of 21st century*, (April):1–16.

Elkington, J. (2001). Enter the Triple Bottom Line. In *The Triple Bottom Line: Does it all Add Up?*, volume 1, pages 1–16.

Endrikat, J., Guenther, E., and Hoppe, H. (2014). Making sense of conflicting empirical findings : A meta-analytic review of the relationship between corporate environmental and financial performance. *European Management Journal*, 32(5):735–751.

Erion, G. (2009). The stock market to the rescue? Carbon disclosure and the future of securities-related climate change litigation. *Review of European Community and International Environmental Law*, 18(2):164–171.

Erlei, M., Leschke, M., and Sauerland, D. (2007). *Neue Institutionenökonomik*. Schäffer Poeschel, Stuttgart, 2. edition.

European Commission (2007). Corporate Social Responsibility (CSR).

European Parliament (2014). Disclosure of non-financial and diversity information by certain large undertaking and groups.

Fabrizi, M., Mallin, C., and Michelon, G. (2014). The Role of CEO's Personal Incentives in Driving Corporate Social Responsibility. *Journal of Business Ethics*, 124(2):311–326.

Fallaha, S., Martineau, G., Bécaert, V., Margni, M., Deschênes, L., Samson, R., and Aoustin, E. (2009). Broadening GHG accounting with LCA: application to a waste management business unit. *Waste Management & Research*, 27(9):885–93.

Fiedler, K. (2007). *Nachhaltigkeitskommunikation in Investor Relations*. Dissertation, Universität Hohenheim.

Fifka, M. S. (2013). Corporate Responsibility Reporting and its Determinants in Comparative Perspective - a Review of the Empirical Literature and a Meta-analysis. *Business Strategy and the Environment*, 22(1):1–35.

Figge, F. and Hahn, T. (2002). *Environmental Sharholder Value Matrix*. Centre for Sustainability Management e.v., Lüneburg.

Fisher-Vanden, K. and Thorburn, K. S. (2011). Voluntary corporate environmental initiatives and shareholder wealth. *Journal of Environmental Economics and Management*, 62(3):430–445.

Fleming, P. (2007). New Consortium Created to Develop Standard Framework for Company Reporting of Climate Risks.

Flick, U. (2007). *Qualitative Sozialforschung. Eine Einführung*.

Font, X., Walmsley, A., Cogotti, S., Mccombes, L., and Häusler, N. (2012). Corporate social responsibility: The disclosure - performance gap. *Tourism Management*, 33(6):1544–1553.

Forster, P. (2016). Case studies for science based targets.

Fowler, S. J. and Hope, C. (2007). A critical review of sustainable business indices and their impact. *Journal of Business Ethics*, 76(3):243–252.

Freedman, M. and Patten, D. M. (2004). Evidence on the pernicious effect of financial report environmental disclosure. *Accounting Forum*, 28(1):27–41.

Freeman, E. R. (1984). *Strategic Management: A Stakeholder Approach*, volume 1.

Freeman, E. R. and Liedtka, J. (1991). corporate social responsibility: a critical approach. *Business Horizons*, (July-August):92–98.

Friedman, M. (1962). Capitalism and Freedom: The Relation Between Economic Freedom and Political Freedom. In *Capitalism and Freedom*, pages 7–17.

Friedman, M. (1970). The Social Responsibility of Business is to Increase its Profits. *The New York Times Magazine*, (32):September 13.

Frooman, J. (1999). Stakeholder Influence Strategies Author. *The Academy of Management Review*, 24(2):191–205.

FTSE (2015a). FTSE4Good Index Series.

FTSE (2015b). Index Inclusion Rules for the Ftse4Good Index Series.

Fürst, E. and Oberhofer, P. (2012). Greening road freight transport: evidence from an empirical project in Austria. *Journal of Cleaner Production*, 33:67–73.

Gerke, W. (2005). Kapitalmärkte - Funktionsweisen, Grenzen, Versagen. In Hungenberg, H. and Meffert, J., editors, *Handbuch Strategisches Management*, pages 215–232. Gabler Verlag, Wiesbaden, 2. auflage edition.

Gibson, K. (2000). The Moral Basis of Stakeholder Theory. *Journal of Business Ethics*, 26(3):245–257.

Glaser, B. G. and Strauss, A. L. (1967). *The Discovery of Grounded Theory: Strategies for Qualitative Research*, volume 1.

Gläser, J. and Laudel, G. (2010). *Experteninterviews und qualitative Inhaltsanalyse - als Instrumente rekonstruierender Untersuchungen*, volume 2., durchg.

Goldsmith, P. D. and Basak, R. (2001). Incentive Contracts and Environmental Performance Indicators. *Environmental and Resource Economics*, (Parkinson 1996):259–279.

Gray, R. (2006). Does sustainability reporting improve corporate behaviour?: Wrong question? Right time? *Accounting and Business Research*, 36(sup1):65–88.

Gray, R. (2010). Is accounting for sustainability actually accounting for sustainability...and how would we know? An exploration of narratives of organisations and the planet. *Accounting, Organizations and Society*, 35(1):47–62.

Gray, R., Kouhy, R., and Lavers, S. (1995). Corporate social and environmental reporting.

Green, J. F. (2010). Private Standards in the Climate Regime: The Greenhouse Gas Protocol. *Business and Politics*, 12(3).

Green, W. and Li, Q. (2012). Evidence of an expectation gap for greenhouse gas emissions assurance. *Accounting, Auditing & Accountability Journal*, 25(1):146–173.

Green, W. and Taylor, S. (2013). Factors that influence perceptions of greenhouse gas assurance provider quality. *International Journal of Auditing*, 17(3):288–307.

Green, W. and Zhou, S. (2013). An international examination of assurance practices on carbon emissions disclosures. *Australian Accounting Review*, 23:54–66.

Green Design Institute (2007). Theory and Method behind EIO-LCA.

Greene, J. C. (2007). *Mixed methods in social inquiry.* Jossey-Bass, San Francisco.

Greene, S. and Lewis, A. (2016). GLEC Framework for Logistics Emissions Methodologies.

Greenhouse Gas Division Environment Canada (2004). Aluminium production - Guidance Manual for Estimating Greenhouse Gas Emissions. Technical Report March, Greenhouse Gas Division Environment Canada, Ottawa.

Greenwald, C. (2014). Financial Materiality of Sustainability. (56):14–18.

Greer, J. and Bruno, K. (1996). *Greenwash: The reality behind corporate environmentalism.* Third World Network.

Greshko, M., Parker, L., and Howard, B. C. (2017). A Running List of How Trump Is Changing the Environment.

Guenther, E. (2009). *Klimawandel und Resilience Management - Interdisziplinäre Konzeption eines entscheidungsorientierten Ansatzes.* Leipzig.

Guenther, E. (2013). *Ökologieorientiertes Management.* Lucius & Lucius, Stuttgart.

Guenther, E., Guenther, T., Schiemann, F., and Weber, G. (2015). Stakeholder Relevance for Reporting: Explanatory Factors of Carbon Disclosure. *Business & Society*, pages 1–37.

Guenther, E. M. and Hoppe, H. (2014). Merging Limited Perspectives. *Journal of Industrial Ecology*, 18(5):689–707.

Guidry, R. P. and Patten, D. M. (2012). Voluntary disclosure theory and financial control variables: An assessment of recent environmental disclosure research. *Accounting Forum*, 36(2):81–90.

Guidry, Ronald P., Patten, D. M. (2010). Market reactions to the first-time issuance of corporate sustainability reports: evidence that quality matters. *Sustainability Accounting Management Policy Journal*, 1(1):33–50.

Hackston, D. and Milne, M. J. (1997). Some determinants of social and environmental disclosures in New Zealand companies. *Accounting Auditing and Accountability Journal*, 9(1):77–108.

Haddock, J. (2006). Consumer influence on internet-based corporate communication of environmental activities: the UK food sector. *British Food Journal*, 107(10):792–805.

Haecker, M., Unser, M., Veeser, A., and Wagner, W. (2001). Investor Relations und Shareholder Value am Neuen Markt. *Finanzbetrieb*, 12:665–678.

Haeseler, H. R. and Hoermann, F. (2006). Wertorientierte Steuerung von Unternehmen und Konzernen mittels Kennzahlen. Beliebte Praktikerkonzepte auf dem wissenschaftlichen Prüfstand. In Seicht, G., editor, *Jahrbuch für Controlling und Rechnungswesen*, pages 115–130. Wien.

Hahn, R. and Kuehnen, M. (2013). Determinants of sustainability reporting: a review of results, trends, theory, and opportunities in an expanding field of research. *Journal of Cleaner Production*, 59:5–21.

Hahn, R. and Luelfs, R. (2013). Legitimizing Negative Aspects in GRI-Oriented Sustainability Reporting: A Qualitative Analysis of Corporate Disclosure Strategies. *Journal of Business Ethics*, pages 1–20.

Hahn, R., Reimsbach, D., and Schiemann, F. (2015). Organizations, Climate Change, and Transparency: Reviewing the Literature on Carbon Disclosure. *Organization & Environment*, 28(1):80–102.

Haigh, M. and Shapiro, M. A. (2011). Carbon reporting: does it matter? *Accounting, Auditing & Accountability Journal*, 25(1):105–125.

Hambrick, D. C. and Mason, P. a. (1984). Echelons : of Reflection The Its Organization as Top a. *Academy of Management Review*, 9(2):193–206.

Harmes, A. (2011). The Limits of Carbon Disclosure: Theorizing the Business Case for Investor Environmentalism. *Global Environmental Politics*, 11(2):98–119.

Hart, S. L. (1995). A Natural-Resource-Based View of the Firm. *Academy of Management Journal*, 20(4):986–1014.

Hart, S. L. and Ahuja, G. (1996). Does it Pay to be Green? An Empirical Examination of the Relationship Between Emission Reduction and Firm Performance. *Business Strategy and the Environment*, 5:30–37.

Hart, S. L. and Dowell, G. (2011). Invited Editorial: A Natural-Resource-Based View of the Firm: Fifteen Years After.

Hart, S. L. and Milstein, M. B. (1999). Global Sustainability and the Creative Destruction of Industries. *Sloan management review*, 41(1):23–33.

Hart, S. L. and Milstein, M. B. (2003). Creating sustainable value. *Academy of Management Executive*, 17(2):56–67.

Hartmann, F., Perego, P., and Young, A. (2013). Carbon accounting: Challenges for research in management control and performance measurement. *Abacus*, 49(4):539–563.

Hawken, P. and Lovins, A. B. (2013). Natural capitalism: the next industrial revolution. In *Technology Review;(USA)*, volume 276, pages 190–212.

Hayek, F. A. (1945). The Use of knowledge in Society. *The American Economic Review*, 35(4):519–530.

Healy, P. M. and Palepu, K. G. (2001). Information asymmetry, corporate disclosure, and the capital markets: A review of the empirical disclosure literature. *Journal of Accounting and Economics*, 31(1-3):405–440.

Heaps, T. A. (2015). What kind of world do you want to invest in?

Hertwich, E., Pease, W., and Koshland, C. (1997). Evaluating the environmental impact of products and production processes : a comparison of six methods. *The Science of Total Environment*, (196):13–29.

Hlavac, M. (2015). stargazer: beatiful LATEX, HTML and ASCII tables from R statistical output.

Hlavac, M., Package, T., Regression, T. W.-f., and Tables, S. S. (2015). Well-Formatted Regression and Summary Statistics Tables.

Hockerts, K., Al, M., Eder-Hansen, J., Krull, P., Midttun, A., Minna, H., Sweet, S., Davidsson, P., Sigurjonsson, T. O., and Nurmi, P. (2009). *CSR-Driven innovation: toward the social purpose business*.

Hoehne, N., Bals, C., Röser, F., Weischer, L., Hagemann, M., Alaoui, A. E., Eckstein, D., Thomä, J., and Rossé, M. (2015a). Developing criteria to align investments with 2 degree C compatible pathways. Technical Report May, New Climate Institue, Germanwatch, 2degree investing initiative, Elmau.

Hoehne, N., Roeser, F., Hagemann, M., Weischer, L., Alaoui, A. E. L., Bals, C., Eckstein, D., Kreft, S., Thomä, J., Rosse, M., and Cochran, I. A. N. (2015b). DEVELOPING 2 degree C-COMPATIBLE INVESTMENT CRITERIA. Technical Report November, New Climate Institue, Germanwatch, 2degree investing initiative.

Hrasky, S. (2012). Visual disclosure strategies adopted by more and less sustainability-driven companies. *Accounting Forum*, 36(3):154–165.

Huang, C.-l. and Kung, F.-h. (2010). Drivers of Environmental Disclosure and Stakeholder Expectation: Evidence from Taiwan. *Journal of Business Ethics*, 96(3):435–451.

Hufschlag, K. (2008). *Informationsversorgung lernender Akteure*.

Hufschlag, K. and Pütz, P. (2013). Carbon Controlling bei Deutsche Post DHL. In Horvarth, P. and Michel, U., editors, *Controlling integriert und global*, pages 217–224. Schäffer Poeschel, Stuttgart.

Indicators, W. G. (2016). Worldwide governance statistics. Technical report, Worldbank.

IPCC (2014). IPCC FIFTH ASSESSMENT REPORT (AR5) "CLIMATE CHANGE 2014" DRAFT SYNTHESIS REPORT Summary for Policymakers provisionally approved. Technical Report October, IPCC Intergovernmental Panel on climate change.

Jacobo, J. (2017). Reactions swift after Trump's withdrawal from Paris Climate Accord.

Jaffe, A. and Peterson, S. (1995). Environmental regulation and the competitiveness of US manufacturing: what does the evidence tell us? *Journal of Economic ...*, 33(1):132–163.

Jira, C. F. and Toffel, M. W. (2013). Engaging Supply Chains in Climate Change. *Manufacturing & Service Operations Management*, 15(4):559–577.

Johnson, B. and Onwuegbuzie, A. J. (2004). Mixed methods research: a research paradigm whose time has come. *Educational Researcher*, 33(7):14–26.

Johnston, I. (2017). China tells Donald Trump there is an 'international responsibility' to act over climate change.

Kaplan, R. S. and Norton, D. P. (1992). The Balanced Scorecard - Measures That Drive Performance. *Harvard Business Review*, 70(1):71–79.

Kasim, T., Baker, R., Cooke, D., and Molloy, J. (2016). Comply or Explain - A review of FTSE 350 companies' environmental reporting and greenhouse gas emission disclosures in annual reports. Technical Report December, CDSB.

Katchova, A. (2013). Panel Data Models.

Kaufmann, D., Kraay, A., and Zoido-Lobatón, P. (1999). Aggregating governance Indicators. *Policy Research Working Paper*, (October 1999):42p.

Kaya, M. (2009). Verfahren der Datenerhebung. In Albers, S., Klapper, D., Konradt, U., Walter, A., and Wolf, J., editors, *Methodik der empirischen Forschung*, chapter 4, pages 49–64. Gabler Verlag, Wiesbaden, 3 edition.

Kennedy, K., Obeiter, M., and Kaufman, N. (2015). Putting a price on carbon: A handbook for U.S. policymakers.

Kim, E.-H. and Lyon, T. (2011). When Does Institutional Investor Activism Increase Shareholder Value? The Carbon Disclosure Project. *The B.E. Jounal of Economic Analysis and Policy*, 11(1).

King, M. E. (2009). King Report on Governance for South Africa 2009. pages 1–142.

Knox-Hayes, J. and Levy, D. (2011). The politics of carbon disclosure as climate governance. *Strategic Organization*, 9:91–99.

Koe, J. and Nga, H. (2009). The influence of ISO 14000 on firm performance. *Social Responsibility Journal*, 5(3):408–422.

Kolk, A., Levy, D., and Pinkse, J. (2008). Corporate Responses in an Emerging Climate Regime: The Institutionalization and Commensuration of Carbon Disclosure. *European Accounting Review*, 17(4):719–745.

KPMG, Fischer, S., Hartmann, T., Hell, C., and Krause, G. (2015). Mehr Wert? Eine Untersuchung von Nutzen und Kosten eines Klimareportings durch deutsche Unternehmen. Technical report, WWF; CDP.

Kräkel, M. (2010). *Organisation und Management*.

Krause, E. and June, M. O. (2015). Carbon Pricing gains popularity with governments, businesses.

Kromrey, H. (2010). *Empirische Sozialforschung*, volume 12. Auflag.

Kuckartz, U. (2014). *Mixed Methods*.

Kuehl, A. (2013). Neues Energielabel für Leuchtmittel.

Kyoto-Protokoll (1997). Protokoll von Kyoto zum Rahmenübereinkommen der Vereinten Nationen über Klimaänderungen. Technical report, Kyoto.

Lackmann, J. (2010). *Die Auswirkungen der Nachhaltigkeitsberichterstattung auf den Kapitalmarkt*. Gabler Verlag.

Lamnek, S. (2010). *Qualitative Sozialforschung*, volume 3. BELTZ, Weinheim, Basel.

Larsen, A. W., Merrild, H., and Christensen, T. H. (2009). Recycling of glass: accounting of greenhouse gases and global warming contributions. *Waste management & research : the journal of the International Solid Wastes and Public Cleansing Association, ISWA*, 27(June):754–762.

Law, S., Weiss, M., Farnworth, E., Bristow, S., Harding, R., Yu, L., Lambin, S., Denner, J., Colman, T., and Rehnborg, M. (2011). The Climate has changed. Technical report, We mean business colalition.

Lee, C. M., Lazarus, M., Smith, G. R., Todd, K., and Weitz, M. (2013). A ton is not always a ton: A road-test of landfill, manure, and afforestation/reforestation offset protocols in the U.S. carbon market. *Environmental Science and Policy*, 33:53–62.

Lee, K.-H. (2012). Carbon accounting for supply chain management in the automobile industry. *Journal of Cleaner Production*, 36:83–93.

Leftwich, R. (1980). Market failure fallacies and accounting information. *Journal of Accounting and Economics*, 2(3):193–211.

Leftwich, R. (1983). Accounting Information in Private Markets: Evidence from Private Lending Agreements. *The Accounting Review*, 58(1):23–42.

Leo, F. D., Vollbracht, M., Brabeck-letmathe, P., Eccles, R. G., and Voigt, T. (2011). Accelerating the Adoption of Integrated Reporting by Robert G. Eccles and George Serafeim. In Leo, F. D. and Vollbracht, M., editors, *CSR Index 2011*, chapter 2.2, pages 70–92. Beirut Boston Pretoria Tianjin Zurich.

Lewis, A., Ehrler, V., Auvinen, H., Maurer, H., Davydenko, I., Burmeister, A., Seidel, S., Lischke, A., and Kiel, J. (2014). Harmonising carbon footprint calculation for freight transport chains. In *Transport Research Arena*, page 10, Paris.

Lewis, J. and Ritchie, J. (2006). Generalising from qualitative Research. In Ritchie, J. and Lewis, J., editors, *Qualitative Research Practice - A guide for social science students and researchers*, chapter Generalisi, pages 263–286.

Liesen, A., Hoepner, A. G. F., Patten, D. M., and Figge, F. (2014). Corporate Disclosure of Greenhouse Gas Emissions in the Context of Stakeholder Pressures: An Empirical Analysis of Reporting Activity and Completeness. *Accounting, Auditing & Accountability Journal*.

Ligges, U. (2006). *Programmieren mit R*. Springer, Berlin, Heidelberg, New York, statistik edition.

limited, S. and Limited, S. (2016). STOXX low carbon indices.

Lucas-Leclin, V. and 2 degree Investing Initiative (2015). Carbon Intensity: different from carbon risk exposure.

Luo, L., Lan, Y. C., and Tang, Q. (2012). Corporate Incentives to Disclose Carbon Information: Evidence from the CDP Global 500 Report. *Journal of International Financial Management and Accounting*, 23(2):93–120.

Luo, L. and Tang, Q. (2014). Does voluntary carbon disclosure reflect underlying carbon performance? *Journal of Contemporary Accounting & Economics*, 10(3):191–205.

Maines, L. A., Bartov, E., Mallett, R., Schrand, C. M., Skinner, D. J., and Vincent, L. (2002). Recommendations on Disclosure of Nonfinanciol Performance Measures. *American Accounting Association*, 16(4):353–362.

Marshall, P. (2009). Climate change and the Great Barrier Reef: impacts and adaptation. Technical report, The Australian Government - Management Ministerial Council commissioned by the Commonwealth Department of Climate Change.

Martinov-Bennie, N. (2012). Greenhouse gas emissions reporting and assurance: reflections on the current state. *Sustainability Accounting, Management and Policy Journal*, 3(2):244–251.

Matisoff, D. C. (2013). Different rays of sunlight: Understanding information disclosure and carbon transparency. *Energy Policy*, 55:579–592.

Matisoff, D. C., Noonan, D. S., and O'Brien, J. J. (2013). Convergence in Environmental Reporting: Assessing the Carbon Disclosure Project. *Business Strategy and the Environment*, 22(5):285–305.

McDonough, W. and Braungart, M. (1998). The next industrial revolution. *The Atlantic*, 282(4):82–92.

McFarland, J. M. (2009). Warming up to climate change risk disclosure. *Fordham Journal of Corporate & Financial Law*, 14:281–323.

McNicholas, P. and Windsor, C. (2011). Can the financialised atmosphere be effectively regulated and accounted for? *Accounting, Auditing & Accountability Journal*, 24:1071–1096.

Mefford, R. N. (2011). The Economic Value of a Sustainable Supply Chain. *Business and Society Review*, 116(1):109–143.

Meuser, M. and Nagel, U. (1991). *ExpertInneninterview - vielfach erprobt, wenig bedacht: ein Beitrag zur qualitativen Methodendiskussion.*

Meuser, M. and Nagel, U. (1994). Expertenwissen und Experteninterview. In *Expertenwissen*, pages 180–192.

Meyer, J. W. and Rowan, B. (1977). Institutionalized organizations: Formal structure as myth and ceremony. *American journal of sociology*, 83(2):340–363.

Michelon, G., Pilonato, S., and Ricceri, F. (2015). CSR reporting practices and the quality of disclosure: An empirical analysis. *Critical Perspectives on Accounting*, 33:59–78.

Miller, R. M. and Plott, C. R. (1985). Product Quality Signaling in Experimental Markets. *Econometrica*, 53(4):837–872.

Milne, M. J. and Grubnic, S. (2011). Climate change accounting research: keeping it interesting and different. *Accounting, Auditing & Accountability Journal*, 24(8):948–977.

Milne, M. J. and Patten, D. M. (2002a). Securing organizational legitimacy. *Accounting, Auditing & Accountability Journal*, 15(3):372–405.

Milne, M. J. and Patten, D. M. (2002b). Securing organizational legitimacy: An experimental decision case examining the impact of environmental disclosures. *Accounting, Auditing & Accountability Journal*, 15(3):372–405.

Mintzberg, H. (1983). the Case for Corporate Social Responsibility. *Journal of Business Strategy*, 4(2):3–15.

Mitchell, R. K., Bradley, A. R. ., and Wood, D. J. (1997). Toward a Theory of Stakeholder Identification and Salience: Defining the Principle of Who and What Really Counts. *The Academy of Management Review*, 22(4):853–886.

Mock, T. J., Rao, S. S., and Srivastava, R. P. (2013). The development of worldwide sustainability reporting assurance. *Australian Accounting Review*, 23(4):280–294.

Murguia, J. M. and Lence, S. H. (2015). Investors' Reaction to Environmental Performance: A Global Perspective of the Newsweek's 'Green Rankings'. *Environmental and Resource Economics*, 60(4):583–605.

Nicholls, M., Faria, P., Labutong, N., Wasmuth, K., Pineda, A. C., Delgado, P., Tornay, C., Huusko, H., Aden, N., and Russel, S. (2015). Mind the Science , Mind the Gap. Technical Report May, We mean business coalition, Paris.

Niehues, N. (2014). Transparenz über CO2-Emissionen in der Logistik als Wettbewerbsvorteil. In Dobersalske, K., Willing, H., and Seeger, N., editors, *Verantwortungsvolles Wirtschaften*, chapter Accounting, pages 265–277. Nomos Verlag, Baden-Baden, 1 edition.

Niehues, N. (2016). Comparability of Carbon figures and the ability of companies to send signals to investors. In *Doctorial colloquium EURAM 2016*, number 1, pages 1–20.

Niehues, N. and Dutzi, A. (2016). The relevance of carbon disclosure – How to disclose in order to achieve comparability in CO2-efficiency. In *Conference proceedings ANZAM*, Brisbane.

Niehues, N. and Dutzi, A. (2018). Disclosing the Invisible – Measurement and Disclosure Pitfalls of CO2 Emissions. In Lindgreen, A., Hirsch, B., Vallaster, C., and Yousafzai, S., editors, *Measuring and controlling sustainability: Spanning Theory and practice*, chapter 16. Francis and Taylor.

Niehues, N., Dutzi, A., and Sawbridge, H. (2017). Voluntary CO2 disclosure and the 2 degree signaling effect. Liverpool. CMS Conference.

Nikolaeva, R. and Bicho, M. (2011). The role of institutional and reputational factors in the voluntary adoption of corporate social responsibility reporting standards. *Journal of the Academy of Marketing Science*, 39(1):136–157.

Oberhofer, P. and Dieplinger, M. (2014). Sustainability in the Transport and Logistics Sector: Lacking Environmental Measures. *Business Strategy and the Environment*, 23(4):236–253.

Oberndorfer, U., Wagner, M., and Ziegler, A. (2011). Does the stock market value the inclusion in a sustainability stock index? An event study analysis for German firms.

O'Dwyer, B., Owen, D., and Unerman, J. (2011). Seeking legitimacy for new assurance forms: The case of assurance on sustainability reporting. *Accounting, Organizations and Society*, 36(1):31–52.

OECD (2015). *The Economic Consequences of Climate Change*. OECD Publishing, Paris.

Okereke, C. (2007). An Exploration of Motivations, Drivers and Barriers to Carbon Management:. *European Management Journal*, 25(6):475–486.

Olson, E. G. (2010). Challenges and opportunities from greenhouse gas emissions reporting and independent auditing.

Parguel, B., Benoît-Moreau, F., and Larceneux, F. (2011). How Sustainability Ratings Might Deter 'Greenwashing': A Closer Look at Ethical Corporate Communication. *Journal of Business Ethics*, 102(1):15–28.

Parmar, B. L., Freeman, E. R., Harrison, J. S., Wicks, A. C., Purnell, L., and de Colle, S. (2010). Stakeholder Theory: The State of the Art. *The Academy of Management Annals*, 4(1):403–445.

Pattberg, P. (2012). How climate change became a business risk: Analyzing nonstate agency in global climate politics. *Environment and Planning C: Government and Policy*, 30(4):613–626.

Patten, D. M. (1991). Exposure, legitimacy, and social disclosure. *Journal of Accounting and Public Policy*, 10(4):297–308.

Patten, D. M. (1992). Intra-industry environmental disclosures in response to the Alaskan oil spill: A note on legitimacy theory. *Accounting, Organizations and Society*, 17(5):471–475.

Patten, D. M. (2002). The relation between environmental performance and environmental disclosure: A research note. *Accounting, Organizations and Society*, 27:763–773.

Patten, D. M. (2015). An insider's reflection on quantitative research in the social and environmental disclosure domain. *Critical Perspectives on Accounting*, 32:45–50.

Peng, R. D. and Matsui, E. (2016). The Art of Data Science: A guide for anyone who works with Data. *Journal of Chemical Information and Modeling*, 53:160.

Pfeffer, J. and Salancik, G. R. (1978). The External Control of Organizations: A Resource Dependence Perspective. *Academy of Management Review*, 4(814):521–532.

Podsakoff, P. M., MacKenzie, S. B., Lee, J.-Y., and Podsakoff, N. P. (2003). Common method biases in behavioral research: a critical review of the literature and recommended remedies. *The Journal of applied psychology*, 88(5):879–903.

Pope Franscesco (2015). *Laudato Si'*.

Popvoich, N. and Schlossberg, T. (2017). How cities and states reacted to Trump's decision to exit the Paris climate deal.

Porter, M. E. (1987). From competitive advantage to corporate strategy. *Harvard Business Review*, 65:43–59.

Porter, M. E. (2016). Do Oil Companies Really Need $4 Billion Per Year of Taxpayers' Money?

Porter, M. E. and Kramer, M. R. (2006). Strategy & society: The link between competitive advantage and corporate social responsibility. *Harvard Business Review*, 84(12):78–92.

Porter, M. E., Linde, C. V. D., and Porter, M. E. (1995). Green and competitive: ending the stalemate. *Harvard Business Review*, 28(6):119–134.

Potter, B., Singh, P. J., and York, J. (2013). Corporate Social Investment Through Integrated Reporting: Critical Issues. In *Seventh Asia Pacific Interdisciplinary Research in Accounting Conference*.

Prado-Lorenzo, J.-M. and Garcia-Sanchez, I.-M. (2010). The Role of the Board of Directors Relevant in Disseminating Gases on Greenhouse Information. *Journal of Business Ethics*, 97(3):391–424.

PriceWatershouseCoopers (2009). Executive guide to King III.

Putt del Pino, S., Levinson, R., and Larsen, J. (2006). *Hot Climate, Cool Commerce: A Service Sector Guide to Greenhouse Gas Management*.

R Development Core Team (2008). *R: A language and environment for statistical computing*. R Foundation for Statistical Computing, Vienna.

Raingold, A. (2010). Carbon Commitments Must Translate into Real Action. *World Economics*, 11(1):181–198.

Ramanathan, K. V. (1976). Toward A Theory of Corporate Social Accounting. *Accounting Review*, 51(3):516–528.

Rankin, M., Windsor, C., and Wahyuni, D. (2011). An investigation of voluntary corporate greenhouse gas emissions reporting in a market governance system: Australian evidence. *Accounting, Auditing & Accountability Journal*, 24(8):1037–1070.

Richards, L. and Morse, J. M. (2013). *Qualitative methods*. Sage.

Richter, R. and Furubotn, E. G. (2003). *Neue Institutionen{ö}konomik: Eine Einf{ü}hrung und kritische W{ü}rdigung*. Mohr Siebeck.

Rikhardsson, P. and Holm, C. (2008). The effect of environmental information on investment allocation decisions - An experimental study. *Business Strategy and the Environment*, 17(6):382–397.

Roberts, L. (2015). Need to get up to speed on Integrated Reporting ? Technical Report January, Accountancy SA.

Robertson, J. and Adani, G. (2016). Adani coal mine: green groups fume over plan for $1b federal loan.

Robinson, M., Kleffner, A., and Bertels, S. (2011). Signaling Sustainability Leadership : Empirical Evidence of the Value of DJSI Membership. *Journal of Business Ethics*, 101:493–505.

Rodgers, W. and Gago, S. (2004). Stakeholder Inf luence on Corporate Strategies Over Time. *Journal of Business Ethics*, 52(4):349–363.

Rodrigue, M. (2014). Contrasting realities: corporate environmental disclosure and stakeholder-released information. *Accounting, Auditing & Accountability Journal*, 27(1):119–149.

Rom, A. and Rohde, C. (2007). Management accounting and integrated information systems: A literature review. *International Journal of Accounting Information Systems*, 8(1):40–68.

RStudio Inc. (2016). R Markdown.

Russo, M. V. and Fouts, P. A. (1997). A Resource-Based Perspective on Corporate Environmental Performance and Profitability. *Academy of Management Journal*, 40(3):534–559.

Saka, C. and Oshika, T. (2014). Disclosure effects, carbon emissions and corporate value. *Sustainability Accounting, Management and Policy Journal*, 5(1):22–45.

San-salvador-del valle, C. and Gómez-bezares, F. (2016). 1542 - DISTRIBUTION OF THE VALUE GENERATED BY THE ECONOMIC ACTIVITY OF AN ORGANIZATION: MODEL AND APPLICATION TO THE COMPANIES IN THE IBEX 35. In *EURAM*, volume 53, pages 1689–1699.

Schaltegger, S. (2011). *Environmental Management Accounting and Supply Chain Management*. Springer, eco-effici edition.

Schaltegger, S. (2013). Die wirtschaftliche Dimension unternehmerischer Nachhaltigkeit. In *Conference proceedings Schmalenbachtagung*, number April, Cologne.

Schaltegger, S. and Burritt, R. (2014). Measuring and Managing Sustainability Performance of Supply Chains. Review and Sustainability Supply Chain Management Framework. *Supply Chain Management: An International Journal*, 19:232–241.

Schaltegger, S., Burritt, R., Zvezdov, D., Hörisch, J., and Tingey-Holyoak, J. (2015). Management Roles and Sustainability Information. Exploring Corporate Practice. *Australian Accounting Review*, 25(4):328–345.

Schmidt, M. (2009). Carbon accounting and carbon footprint - more than just diced results? *International Journal of Climate Change Strategies and Management*, 1(1):19–30.

Schmidt, M. (2012a). Integrierte Berichterstattung - Weckruf für die Finanzberichterstattung. *IRZ: Zeitschrift für Internationale Rechnungslegung*, 7(4):137–138.

Schmidt, M. (2012b). *Moeglichkeiten und Grenzen einer integrierten Finanz- und Nachhaltigkeitsberichterstattung*. Baetge, J{ö}rg; Kirsch, Hans-J{ü}rgen, Duesseldorf, 1. auflage edition.

Schnell, C. (2015). Der Absturz vom Olymp.

Scholes, E. and Clutterbuck, D. (1998). Communication with stakeholders: An integrated approach. *Long Range Planning*, 31(2):227–238.

Searcy, C. and Elkhawas, D. (2012). Corporate sustainability ratings: An investigation into how corporations use the Dow Jones Sustainability Index. *Journal of Cleaner Production*, 35:79–92.

Senge, P. M. and Carstedt, G. (2001). Innovating Our Way to the Next Industrial Revolution. *MIT Sloan Management Review*, 42(2):24–38.

Sharma, S. and Henriques, I. (2005). Stakeholder influences on sustainability practices in the Canadian forest products industry. *Strategic Management Journal*, 26(2):159–180.

Sharma, S. and Vredenburg, H. (1998). Proactive corporate environmental strategy and the development of competitively valuable organizational capabilities. *Strategic Management Journal*, 19:729.

Shocker, A. D. and Sethi, S. P. (1973). An Approach to Incorporating Societal Preferences in Developing Corporate Action Strategies. *California Management Review*, 15(4):97–105.

Shrivastava, P. (1995). Environmental technologies and competitive advantage. *Strategic Management Journal*, 16(Summer):183–200.

Silverman, D. (2006). *Interpreting Qualitative Data - Methods for Analysing Talk, Text and Interaction*. Sage Publications, London, Thousand Oaks, New Dehli, 2nd edition.

Skelton, A. (2013). EU corporate action as a driver for global emissions abatement: A structural analysis of EU international supply chain carbon dioxide emissions. *Global Environmental Change*, 23(6):1795–1806.

Skinner, D. J. (1993). The investment opportunity set and accounting procedure choice. *Journal of Accounting and Economics*, 16(4):407–445.

Solomon, J. F., Solomon, A., Norton, S. D., and Joseph, N. L. (2011). Private climate change reporting: an emerging discourse of risk and opportunity? *Accounting, Auditing & Accountability Journal*, 24(8):1119–1148.

South Africa National Treasury (2013). Carbon Tax Policy Paper. Technical Report May, South Africa National Treasury, Johannesbourg.

Spence, M. (1973). Job Market Signaling. *Quarterly Journal of Economics*, 87:355–374.

Stanny, E. (2013). Voluntary Disclosures of Emissions by US Firms. *Business Strategy and the Environment*, 22(3):145–158.

Stanny, E. and Ely, K. (2008). Corporate environmental disclosures about the effects of climate change. *Corporate Social Responsibility and Environmental Management*, 15(6):338–348.

Stawinoga, M. (2012). *Nachhaltigkeitsberichterstattung im Lagebericht - Konzeptionelle und empirische Analyse einer integrierten Berichterstattung*. Erich Schmidt Verlag, Hamburg, 1. auflage edition.

Stechemesser, K. and Guenther, E. (2012). Carbon accounting: A systematic literature review. *Journal of Cleaner Production*, 36:17–38.

Stein, M. and Khare, A. (2009). Calculating the carbon fottprint of a chemical plant: A case study for Akzonobel. *Journal of Environmental Assessment Policy and Management*, 11(03):291–310.

Stern, N. (2006). *The Economics of Climate Change*, volume 7. Cambridge.

Struebing, J. (2004). *Grounded Theory. Zur sozialtheoretischen und epistemologischen Fundierung des Verfahrens der empirisch begründeten Theoriebildung*, volume 15. Springer VS.

Stubbs, W. and Higgins, C. (2014). Integrated Reporting and internal mechanisms of change. *Accounting, Auditing & Accountability Journal*, 27(7):1068–1089.

Suchman, M. C. (1995). Managing Legitimacy: Strategic and Approaches. *Academy of Management Review*, 20(3):571–610.

Sullivan, R. and Gouldson, A. (2012). Does voluntary carbon reporting meet investors' needs? *Journal of Cleaner Production*, 36:60–67.

Talbot, D. and Boiral, O. (2013). Can we trust corporates GHG inventories? An investigation among Canada's large final emitters. *Energy Policy*, 63:1075–1085.

Tcholak-Antitch, Z., Carnac, T., Harris, C., Gilbertson, S., Baker, B., Banks, M., Gerholdt, R., Leonard, L., Patterson, D., Sobel, A., Schallert, B., Spitzer, M., WWF, CDP, Tcholak-Antitch, Z., Carnac, T., Harris, C., Gilbertson, S., Baker, B., Banks, M., Gerholdt, R., Leonard, L., Patterson, D., Sobel, A., Schallert, B., and Spitzer, M. (2013). The 3% solution. Technical report, WWF, CDP, Washington, New York.

Teddlie, C. and Tashakkori, A. (2003). Major Issues and Controversies in the Use of Mixed Methods in the Social and Behavioural Sciences. In *Handbook of Mixed Methods in Social & Behavioral Research*, pages 3–50.

Teng, M.-j., Wu, S.-y., and Chou, S. J.-h. (2014). Environmental Commitment and Economic Performance - Short-Term Pain for Long-Term Gain. *Environmental Policy and Governance*, 27(December 2013):16–27.

The Carbon Disclosure Project (2015). CDP Global Climate Change Report 2015. Technical Report October.

Thijssens, T., Bollen, L., and Hassink, H. (2015). Secondary Stakeholder Influence on CSR Disclosure: An Application of Stakeholder Salience Theory.

Thomae, J., Weber, C., and Dupre, S. (2016). Measuring Progress on Greening Financial Markets. Technical report, 2degreeInvestingInititative, Paris, New York, London.

Thomson, I. (2015). 'But does sustainability need capitalism or an integrated report' a commentary on 'The International Integrated Reporting Council: A story of failure' by Flower, J. *Critical Perspectives on Accounting*, 27:18–22.

TNO, de Ree, D., Ton, J., Davydenko, I., Chen, M., Kiel, J., Auvinen, H., and Mäkelä, K. (2012). D3.1 Assessment and typology of existing CO2 calculation tools and methodologies. Technical report, COFRET Carbon Footprint of Freight Transport.

Torres-reyna, O. (2010). Getting Started in Fixed/Random Effekts Models using R.

Trueman, B. (1986). Why do managers voluntarily release earnings forecasts? *Journal of Accounting and Economics*, 8(1):53–71.

Tuppen, C. (2008). Climate Stabilisation Intensity Targets - A new approach to setting corporate climate change targets. page 13.

Valor, C. (2008). Can consumers buy responsibly? Analysis and solutions for market failures. *Journal of Consumer Policy*, 31(3):315–326.

Velte, P. (2008). *Intangible Assets und Goodwill im Spannungsfeld zwischen Entscheidungsrelevanz und Verlässlichkeit*. Hamburg.

Verhoef, E., Nijkamp, P., and Rietveld, P. (1996). THE TRADE-OFF BETWEEN EFFICIENCY, EFFECTIVENESS, AND SOCIAL FEASIBILITY OF REGULATING ROAD TRANSPORT EXTERNALITIES. *Transportation Planning and Technology*, 19:247–263.

Verrecchia, R. E. (1983). Discretionary disclosure. *Journal of Accounting and Economics*, 5:179–194.

Verrecchia, R. E. (1999). Disclosure and the cost of capital: A discussion. *Journal of Accounting and Economics*, 26(1-3):271–283.

Vlugt, I. v. d. V., Wber, C. L., Matthews, S., Sawbridge, H., and Griffin, P. (2015). Modeling of Corporate GHG Emissions Methodology. Technical report, CDP.

Wackernagel, M. and Rees, W. E. (1997). Perceptual and structural barriers to investing in natural capital: Economics from an ecological footprint perspective. *Ecological Economics*, 20(1):3–24.

Wagenhofer, A. (1990a). *Informationspolitik im Jahresabschluss - Freiwillige Informationen und strategische Bilanzanalyse.* Physica Verlag, Heidelberg.

Wagenhofer, A. (1990b). Voluntary disclosure with a strategic opponent. *Journal of Accounting and Economics*, 12(4):341–363.

Wagner, M. (2010). The role of corporate sustainability performance for economic performance: A firm-level analysis of moderation effects. *Ecological Economics*, 69(7):1553–1560.

Walley, N. and Whitehead, B. (1994). It's not easy being green. *Business and the Environment*, 36:81.

Walls, J. L., Phan, P. H., and Berrone, P. (2011). *Measuring Environmental Strategy: Construct Development, Reliability, and Validity*, volume 50.

Walther, G., Engel, B., and Spengler, T. (2009). Bewertung und Verbesserung kumulierter Emissionsintensitäten. pages 191–199.

Watkins, D. C. and Gioia, D. (2015). *Mixed Methods Research.* Oxford University Press.

Watts, R. L. and Zimmerman, J. L. (1986). Positive accounting theory.

Wegener, M., Elayan, F. a., Felton, S., and Li, J. (2013). Factors Influencing Corporate Environmental Disclosures. *Accounting Perspectives*, 12(1):53–73.

Weidema, B. P., Thrane, M., Christensen, P., Schmidt, J., and Løkke, S. (2008). Carbon Footprint:A Catalyst for Life Cycle Assessment? *Journal of Industrial Ecology*, 12(1):1–7.

Weischer, L., Warland, L., Eckstein, D., Hoch, S., Michaelowa, A., and Koehler, M. (2016). Investing in Ambition. Technical report, Germanwatch, Perspectives Climate Group GmbH, Bonn, Freiburg.

Weißenberger, B. E. (2014). Integrated Reporting: Fragen (und Antworten) aus der Diskussion um die integrierte Rechnungslegung. *Controlling - Zeitschrift für erfolgsorientierte Unternehmenssteuerung*, 26(8/9):440–446.

Wickham, H. and Francois, R. (2016). A Grammar of Data Manipulation.

Windolph, S. E. (2013). Motivations, Organizational Units, and Management Tools. Taking Stock of the Why, Who, and How of Implementing Corporate Sustainability Management. (August):157.

Witzel, A. (1982). *Verfahren der qualitativen Sozialforschung: Überblick und Alternativen*, volume 322. Campus Verlag.

Wong, R. and Millington, A. (2014). Corporate social disclosures: a user perspective on assurance. *Accounting, Auditing & Accountability Journal*, 27(5):863–887.

Wright, P. and Ferris, S. P. (1997). Research Notes and Communications Agency Conflict and Corporate Strategy: the Effect of Divestment on Corporate Value. *Strategic Management Journal*, 18(November 1994):77–83.

Xuchao, W., Priyadarsini, R., and Siew Eang, L. (2010). Benchmarking energy use and greenhouse gas emissions in Singapore's hotel industry. *Energy Policy*, 38(8):4520–4527.

Yunus, S., Elijido-Ten, E., and Abhayawansa, S. (2016). Determinants of carbon management strategy adoption: Evidence from Australia's top 200 publicly listed firms. *Managerial Auditing Journal*, 31(2):156–179.

Zvezdov, D. (2011). Rolling out Corporate Sustainability Accounting : A Set of Challenges. *Journal of Environmental Sustainability*, 2:19–28.

Zvezdov, D. D. and Schaltegger, S. (2014). Carbon Management Accounting: A Systematic Literature Review. In *EMAN Conference*, Rotterdam.

Chapter 6: Appendix

	country	N	in %	Mean	SD	Min	Median	Max
1	USA	255	20.00	33	75	0.09	8	430
2	Japan	168	13.00	8	35	0.13	1	330
3	UK	124	10.00	30	62	0.04	8	354
4	Canada	73	6.00	27	55	0.26	6	295
5	China	59	5.00	32	64	0.26	14	427
6	South Korea	52	4.00	53	103	0.09	21	448
7	Germany	51	4.00	21	30	0.26	15	193
8	France	46	4.00	57	93	0.26	23	386
9	Switzerland	45	3.00	23	40	0.18	8	221
10	South Africa	43	3.00	66	103	0.53	38	497
11	Sweden	37	3.00	38	82	0.18	12	355
12	India	36	3.00	3	6	0.13	1	25
13	Brazil	35	3.00	37	67	0.1	20	404
14	Italy	31	2.00	38	66	0.62	18	282
15	Spain	31	2.00	26	17	0.76	25	60
16	Finland	24	2.00	29	26	0.95	21	118
17	Norway	21	2.00	30	50	0.22	14	211
18	Turkey	20	2.00	39	106	0.44	7	482
19	Netherlands	18	1.00	29	59	1.73	14	258
20	Denmark	15	1.00	63	113	0.38	15	435
21	Ireland	13	1.00	16	13	1.58	12	39
22	Colombia	11	1.00	17	14	0.79	15	46
23	Austria	10	1.00	42	108	0.26	7	350
24	Belgium	8	1.00	22	16	0.72	20	48
25	New Zealand	8	1.00	10	15	0.79	6	45
26	Singapore	8	1.00	18	26	0.21	4	67
27	Thailand	8	1.00	12	12	0.74	6	32
28	Russia	7	1.00	6	6	0.2	7	15
29	Portugal	5	0.00	59	49	29.28	42	146
30	Greece	4	0.00	34	29	7.48	31	64
31	Israel	4	0.00	19	18	0.18	18	42
32	Mexico	4	0.00	46	42	0.39	42	98

33	Poland	4	0.00	9	8	0.26	9	16
34	Bermuda	3	0.00	23	18	5.59	22	41
35	Chile	3	0.00	18	28	1.71	2	50
36	Philippines	3	0.00	3	2	1.31	2	5
37	Hungary	2	0.00	34	37	7.6	34	60
38	Indonesia	2	0.00	1	0	0.71	1	1
39	Luxembourg	2	0.00	23	3	20.79	23	26
40	Malaysia	2	0.00	28	3	26.32	28	30
41	Peru	2	0.00	3	2	1.51	3	4
42	United Arab Emirates	2	0.00	11	14	0.86	11	21
43	Cyprus	1	0.00	3	NA	3.45	3	3
44	Guernsey	1	0.00	14	NA	13.76	14	14

Table 6.1: Reporting principles by country mapping

Question	Detailed questions
Introduction	Small talk to ensure trustful atmosphere, explain interview, get missing master data (e.g. Name, function, company, branch).
Expert validity test	Member of initiatives (e.g. CDP, regulation initiative, industry initiative or other expert group). Number of years working in the area of environmental reporting / analyzing environmental reports. Percentage of the working time related to CO_2 reporting/analyzes.
Main question	How can I see, if a company is CO_2 efficient compared to its peer (industry) companies?
What would you focus on, to make CO_2 emissions comparable?	Which factors limit the comparability of CO_2 emissions numbers? What is the role of the standard? Scopes? Dept of value added? Methods used? Data collection process (Process, Audit, IT-System, Responsibility in company, etc.)
Could you estimate the reporting error in your analyzed/reported CO_2 emissions?	What is the competitors' value? How did you estimate it? Do you have a quality assurance process? How does it work?
Which criteria are used by investors/stakeholders to analyze CO_2 efficiency?	Which factors limit the comparability of CO_2 emissions numbers? What is the role of the standard? Scopes? Dept of value added? Methods used? Data collection process (Process, Audit, IT-System, Responsibility in company, etc.)
Verification of the sample	In what areas would you search for experts regarding CO_2 emissions-reporting?
Snowball system	What experts would you recommend personally for these types of interviews?

Table 6.2: Interview guideline for qualitative study

	Short name	Long name	technical name
H1	DepVar	Reporting principles (log scale)	log_reporting_principles
H3	DepVar	Auditscore for own CO_2 emissions	S12auditscore
2	P1	Compliant - 3-Degree target	year3dgcompl
3	P2	Compliant - Climate Stabilisation Intensity	NoCSIcompliantyears
4	P3	Self-reported trend	S12performance
5	P4	Emission-reduction initiatives	EmissionReductionInitiatives
6	P5	Downstream reduction initiatives	DownstreamReduction
7	P6	Target category	TargetCategory
8	P7	Detail Abs Target	DetailAbsTarget
9	M1	GHG politics	GHGPolitics
10	M2	Legal system	legalsystem
11	M3	Public voice	PubVoice
12	M4	Government effectiveness	GovEff
13	M5	Intensity factor	IntensityFactor
14	M6	Revenue (log scale)	log_revenue
15	M7	GRI index listing	investor
16	M8	Significant CDP	significantCDP
17	M9	Significant GRI	significantGRI
18	M10	Significant GICS	significantGICS
19	M11	Significant country	significantcountry

Table 6.3: Overview of the variables with short, long and technical names

	H1	H2	P1	P2	P3	P4	P5	P6	P7	M1
H1	1.00	0.70	0.26	0.26	-0.04	0.52	0.43	0.49	-0.12	0.16
H2	0.70	1.00	0.17	0.14	-0.02	0.48	0.38	0.49	-0.15	0.14
P1	0.26	0.17	1.00	0.59	0.02	0.15	0.15	0.16	-0.12	0.01
P2	0.26	0.14	0.59	1.00	-0.03	0.11	0.13	0.12	-0.11	0.04
P3	-0.04	-0.02	0.02	-0.03	1.00	0.01	-0.01	-0.09	0.07	-0.06
P4	0.52	0.48	0.15	0.11	0.01	1.00	0.58	0.58	-0.00	0.29
P5	0.43	0.38	0.15	0.13	-0.01	0.58	1.00	0.45	-0.04	0.21
P6	0.49	0.49	0.16	0.12	-0.09	0.58	0.45	1.00	-0.44	0.19
P7	-0.12	-0.15	-0.12	-0.11	0.07	-0.00	-0.04	-0.44	1.00	0.06
M1	0.16	0.14	0.01	0.04	-0.06	0.29	0.21	0.19	0.06	1.00
M2	-0.00	-0.07	0.02	0.08	0.02	0.15	0.03	-0.02	0.11	0.22
M3	-0.01	-0.12	0.09	0.09	-0.07	-0.03	0.01	-0.03	0.03	0.27
M4	-0.08	-0.17	0.07	0.09	-0.02	-0.09	-0.05	-0.07	-0.01	0.00
M5	0.06	0.10	0.03	-0.02	0.21	0.05	0.07	0.05	0.03	-0.07
M6	-0.00	0.03	0.09	0.09	0.03	-0.25	-0.05	-0.07	-0.22	-0.55
M7	0.22	0.19	0.20	0.21	0.07	0.13	0.16	0.18	-0.12	0.10
M8	0.06	0.07	0.01	-0.02	0.04	0.00	0.02	0.04	-0.03	-0.04
M9	0.01	-0.03	-0.00	-0.01	0.02	-0.09	-0.13	-0.07	0.01	-0.09
M10	0.05	0.05	0.02	-0.02	0.06	-0.01	0.18	0.04	-0.04	-0.05
M11	0.19	0.23	-0.03	-0.02	0.04	0.16	0.09	0.08	0.01	-0.06

Table 6.4: Correlation table 1/2

	M2	M3	M4	M5	M6	M7	M8	M9	M10	M11
H1	-0.00	-0.01	-0.08	0.06	-0.00	0.22	0.06	0.01	0.05	0.19
H2	-0.07	-0.12	-0.17	0.10	0.03	0.19	0.07	-0.03	0.05	0.23
P1	0.02	0.09	0.07	0.03	0.09	0.20	0.01	-0.00	0.02	-0.03
P2	0.08	0.09	0.09	-0.02	0.09	0.21	-0.02	-0.01	-0.02	-0.02
P3	0.02	-0.07	-0.02	0.21	0.03	0.07	0.04	0.02	0.06	0.04
P4	0.15	-0.03	-0.09	0.05	-0.25	0.13	0.00	-0.09	-0.01	0.16
P5	0.03	0.01	-0.05	0.07	-0.05	0.16	0.02	-0.13	0.18	0.09
P6	-0.02	-0.03	-0.07	0.05	-0.07	0.18	0.04	-0.07	0.04	0.08
P7	0.11	0.03	-0.01	0.03	-0.22	-0.12	-0.03	0.01	-0.04	0.01
M1	0.22	0.27	0.00	-0.07	-0.55	0.10	-0.04	-0.09	-0.05	-0.06
M2	1.00	0.27	0.28	-0.03	-0.40	0.22	-0.03	0.03	-0.06	-0.39
M3	0.27	1.00	0.73	-0.03	-0.29	0.01	-0.01	-0.14	-0.08	-0.49
M4	0.28	0.73	1.00	-0.10	-0.14	0.01	0.01	-0.02	-0.08	-0.42
M5	-0.03	-0.03	-0.10	1.00	0.00	0.25	-0.02	-0.09	0.18	0.01
M6	-0.40	-0.29	-0.14	0.00	1.00	0.22	0.01	0.01	0.02	0.01
M7	0.22	0.01	0.01	0.25	0.22	1.00	0.01	0.01	0.02	0.17
M8	-0.03	-0.01	0.01	-0.02	0.01	-0.06	1.00	0.00	-0.08	-0.05
M9	0.03	-0.14	-0.02	-0.09	0.01	-0.04	0.00	1.00	0.02	0.04
M10	-0.06	-0.08	-0.08	0.18	0.02	-0.08	0.02	0.10	1.00	0.09
M11	-0.39	-0.49	-0.42	0.01	0.17	-0.05	0.04	0.09	0.07	1.00

Table 6.5: Correlation table 2/2

Dependent variable:

| | S12auditscore | | | log_reporting_principles |
	Full data	Highend	Tail	Hypothesis 1
Constant	0.740*** (0.139)	1.414*** (0.256)	0.487*** (0.131)	−0.001 (0.001)
s12performance	−0.001 (0.001)	−0.0004 (0.002)	−0.002 (0.001)	
year3dgcompl	0.012* (0.006)	0.002 (0.011)	0.001 (0.006)	
NoCSIcompliantyears	0.010 (0.006)	−0.013 (0.011)	0.007 (0.006)	
DownstreamReduction	0.442*** (0.056)	0.066 (0.095)	0.287*** (0.061)	0.369*** (0.042)
EmissionReductionInitiatives	1.628*** (0.066)	3.378*** (0.112)	1.213*** (0.071)	0.948*** (0.051)
TargetCategory	0.615*** (0.036)	0.232*** (0.051)	0.687*** (0.041)	0.323*** (0.027)
DetailAbsTarget	−0.013 (0.011)	0.055*** (0.019)	−0.009 (0.012)	0.007 (0.008)
GHGPolitics	0.001 (0.002)	−0.004 (0.003)	0.003* (0.001)	0.054 (0.085)
legalsystem	−0.068** (0.033)	−0.035 (0.057)	−0.056* (0.031)	
PubVoice	0.060* (0.033)	0.051 (0.066)	0.004 (0.031)	0.154 (0.188)
GovEff	−0.098*** (0.033)	−0.109* (0.060)	−0.034 (0.031)	0.092 (0.097)
log_revenue	0.001 (0.007)	−0.021* (0.013)	0.013** (0.006)	−0.121** (0.049)
IntensityFactor	0.009 (0.030)	−0.048 (0.051)	0.028 (0.028)	
investor	0.001** (0.001)	0.0004 (0.001)	0.001 (0.001)	

Note: *p<0.1; **p<0.05; ***p<0.01

Table 6.6: Hypothesis 3 tested with panel data for years 2011 - 2015 different disclosure sets 1/2

Dependent variable:

	S12auditscore			log_reporting_principles
	Full data	Highend	Tail	Hypothesis 1
significantCDP	0.367** (0.151)	0.309 (0.227)	0.251 (0.160)	
significantGRI	−0.022 (0.042)	−0.020 (0.071)	−0.074* (0.042)	
significantGICS	0.029 (0.051)	0.044 (0.090)	0.022 (0.048)	
significantcountry	0.106*** (0.037)	0.048 (0.064)	0.074** (0.036)	
Constant	0.740*** (0.139)	1.414*** (0.256)	0.487*** (0.131)	
Observations	5,897	1,477	4,420	4,579
R^2	0.654	0.870	0.565	0.572
Adjusted R^2	0.653	0.869	0.563	0.571
F Statistic	616.170*** (df = 18; 5878)	543.235*** (df = 18; 1458)	317.104*** (df = 18; 4401)	677.252*** (df = 9; 4569)

Note: *p<0.1; **p<0.05; ***p<0.01

Table 6.7: Hypothesis 3 tested with panel data for years 2011 - 2015 different disclosure sets 2/2

	Dependent variable:	
	H1-2015	H3-2015
GHG politics	0.002 (0.004)	0.017** (0.007)
Legal system	−0.040 (0.074)	−0.280** (0.142)
Public voice	0.253*** (0.066)	0.318** (0.127)
Government effectiveness	−0.168** (0.067)	−0.309*** (0.129)
Intensity factor	0.008 (0.066)	0.087 (0.128)
Revenue (log scale)	0.035** (0.015)	0.071** (0.029)
GRI index listing	0.004*** (0.001)	0.009*** (0.002)
Significant CDP	0.890*** (0.331)	1.500** (0.639)
Significant GRI	0.326*** (0.092)	0.470*** (0.178)
Significant GICS	0.122 (0.114)	0.182 (0.219)
Significant country	0.401*** (0.082)	0.722*** (0.159)
Observations	1,301	1,301
R^2	0.422	0.402
Adjusted R^2	0.414	0.394
Residual Std. Error (df = 1282)	0.983	1.895
F Statistic (df = 18; 1282)	51.946***	47.931***
Note:		$^*p<0.1;$ $^{**}p<0.05;$ $^{***}p<0.01$

Table 6.8: Hypothesis 4 overview of stakeholder variables, single-year view

	Dependent variable:	
	log_reporting_principles H1-Panel	S12auditscore H3-Panel
GHG politics	0.054 (0.085)	0.001 (0.002)
Legal system		-0.068^{**} (0.033)
Public voice	0.154 (0.188)	0.060^{*} (0.033)
Government effectiveness	0.092 (0.097)	-0.098^{***} (0.033)
Intensity factor	-0.121^{**} (0.049)	0.001 (0.007)
Revenue (log scale)		0.009 (0.030)
GRI index listing		0.001^{**} (0.001)
Significant CDP		0.367^{**} (0.151)
Significant GRI		-0.022 (0.042)
Significant GICS		0.029 (0.051)
Significant country		0.106^{***} (0.037)
Observations	4,579	5,897
R^2	0.572	0.654
Adjusted R^2	0.571	0.653
F Statistic	677.252^{***} (df = 9; 4569)	616.170^{***} (df = 18; 5878)
Note:		$^{*}p<0.1$; $^{**}p<0.05$; $^{***}p<0.01$

Table 6.9: Hypothesis 4 overview panel view